The Replacement Trap

A Field Manual for Human Quality Control in an AI Workplace

Renae C. Linde

ISBN: 979-89930830-4-9

Independent publication by CRF Luttrell

Contents

Introduction

Every shift, workers across every major industry receive paperwork that came from somewhere upstream. A warehouse receiver reads a dispatch manifest. A clinical support worker reads a chart summary. A field technician reads a diagnostic report. A billing clerk reads an invoice payload. Each person reads the document, trusts what it says, and acts on it. But the problem is common: none of them can easily tell who wrote it.

This problem has grown, and it now carries a much higher cost. Today's workplaces use automated tools that can write work documents and instruction sheets faster than any human team could ever match. These tools produce polished, professional-looking text: it reads cleanly, carries no formatting errors, and gives no sign of where it came from. An AI system can generate a document that looks just like something a senior colleague typed that same morning, so frontline workers who receive both may have no easy way to tell one from the other unless the workplace already labels its sources.

This condition has a name: source opacity. It means a worker has to do a task using material whose reliability they can't confirm on their own. No one can trace this work back to where it started, and that makes it a real workplace hazard. It doesn't just live in a server room or a strategy document. It lands on a clipboard or a loading dock every morning.

The tools that produce these documents run on one simple rule: they calculate which word, phrase, or figure is most likely to come next, based on patterns learned from old data. They aren't searching for truth, and they don't check facts against independent sources. Because these tools are so good at building smooth, well-ordered sentences, the output sounds confident, but a confident tone doesn't mean the content is accurate. When these systems train again and again on their own output, instead of on records written by real people, their accuracy degrades even as the text stays smooth and easy to read. That's the trap: the document that looks the most polished may be hiding the deepest mistake.

Top managers at many companies often judge these tools by cost dashboards and speed numbers. When a software vendor shows that a document tool cuts office labor hours by some measurable percentage, that figure can help win the sale. What many dashboards leave out is the cleanup work frontline workers do later, if the automated output turns out to contain errors. Automated systems can also erode institutional knowledge, cutting the feedback loop between frontline workers and the data systems their companies depend on, and that loss often doesn't show up as a line-item cost. The worker who catches a mistake and fixes it, before a shipment ships wrong or a patient receives the wrong drug, often gets no credit from the company for the save. The catch stays invisible, and the speed metric stays exactly the same.

This gap drives a real fear among frontline workers. They watch their companies buy tools sold as labor-saving systems, and they read an implied message: fewer workers, at lower cost. That fear of replacement isn't foolish; it follows the company's own logic. But the fear rests on a flawed idea. Automated tools produce work that still needs human judgment to guide it. These systems confabulate: they generate words that sound right but aren't true. A tool that produces believable mistakes at high speed does not replace the worker who catches those mistakes. It just adds to that worker's mental load, while the company rarely acknowledges that the load exists at all.

The real threat runs deeper than any single tool. When a company demands speed but sets no rules for who is accountable, workers end up acting on unverified content under pace pressure and constant electronic monitoring. When institutional memory erodes, because automated systems overwrite local know-how with no provenance tracking, companies lose the expertise that kept the work running. And when workers are judged on speed instead of accuracy, the reward system ends up punishing the very judgment calls that prevent serious failures. The gospel of speed, applied without verification discipline, turns daily work into a race against being replaced by machines, rather than a system workers can govern and survive.

The nine skills in this book meet that reality head-on. They apply in workplaces where incoming documents can't always be verified at a glance, and where pace pressure discourages workers from stopping to check them. The skills are manual verification protocols, error logging procedures, localized documentation systems, pre-flight checks, handoff verification buffers, the format bridge, the exception flag system, collaborative audits, and informed refusal protocols. Each skill targets one point in the daily workflow where unverified content can cause downstream harm. Together, they form one complete system that workers can apply step by step, across most trades and tool sets.

These skills all run on a single rule: treat every incoming claim as unverified until an independent check confirms it, regardless of who or what produced it. That rule doesn't slow down real work. A skilled worker can run most of these checks inside the gaps that already exist between receiving a document and acting on it. Many of these skills need no green light from the company to get started; a worker can put them to use alone, though a few benefit from organizational buy-in. It only takes a worker who understands how automated tools generate believable mistakes, and who chooses to govern those tools instead of obeying them without question.

The nine skills in this book make that choice to govern practical and doable on most shifts, in most workplaces, on any floor where work documents can't always be verified at a glance.

Skill 1: Manual Verification Protocols

Operating Under Source Opacity

The incoming dispatch manifest arrives formatted cleanly. The column headers align. The product codes follow a logical sequence. The quantities read as precise figures with no visible errors in spelling or syntax. Nothing about the document signals a problem. But that clean formatting tells a warehouse receiver nothing about whether the data underneath it is correct.

This is the core problem that source opacity creates. Over years on the job, workers in many industries have learned to trust a well-formatted document. A typed work order looked more trustworthy than a handwritten note. A printed label looked more official than a pencil mark. That made sense back when a careful person was behind the formatting. Automated tools have cut that link completely. An automated text tool turns out the same clean formatting whether the facts are right or wrong. The system cannot tell a correct answer from a wrong one that merely sounds correct. It produces both with the exact same polish.

The National Institute of Standards and Technology (NIST) identified this property directly in its AI Risk Management Framework, noting that AI systems can generate outputs that appear confident and well-structured while containing factually incorrect claims (Tabassi, 2023). That finding matters directly on the warehouse floor, and in workplaces like it. A worker reading a document cannot see the process that made it. The worker sees only the finished surface, and that surface reveals nothing.

Source labeling makes the problem worse instead of solving it. When companies route AI-generated documents through the same systems and approval chains used for documents written by people, workers may have no easy way to tell the two apart just by looking. A generated work summary enters the same inbox as a supervisor's typed instructions. A system-built chart update sits in the same folder as a clinician's hand-checked note. Workers often have no reliable way to sort the two by sight. Even when a document carries a system-generated header or a software timestamp, that marker only proves a tool touched the document. It does not prove that any human reviewed the content before it reached the floor.

This creates a specific kind of failure when workers try to check one document against another digital source. The warehouse receiver shows most clearly how this failure works. A receiver's normal practice is to match an incoming dispatch manifest against the inventory system's live network feed. The logic looks sound: two data points checked against each other should show up any mismatch. But the logic fails the moment the same upstream system that wrote the manifest also fed the network feed. If an AI-assisted procurement tool wrote the manifest and updated the network feed in the same automated pass, both records carry the same error at once. The receiver compares manifest to feed and finds agreement. The shipment clears. The error slips through unchallenged.

Shumailov and colleagues demonstrated in 2024 that when AI systems generate outputs that then feed into subsequent AI-assisted processes, errors propagate and compound across the chain rather than canceling each other out (Shumailov et al., 2024). In a warehouse, that finding plays out as a real, specific chain of events. The procurement tool generates a manifest with an incorrect unit count. The inventory system ingests the manifest and updates its own records. The receiver checks the manifest against the inventory system. Both documents agree. Both documents are wrong. The physical pallet sitting on the dock holds a different quantity than either record shows.

The physical container tag breaks that error chain because it has no connection to the upstream digital system at all. The tag shows what a human actually counted and wrote down at the point of origin. That count might contain its own human error, but it comes from a separate physical process, outside the automated chain. When the receiver checks the manifest against the tag and finds a mismatch, the mismatch shows up because two independent sources now disagree. That disagreement tells the receiver something useful: at least one record needs a closer look before the shipment clears.

This is the logic behind requiring an independent, uncorrupted reference source for verification, no matter who or what wrote the incoming document. Independence means the reference source has no data link to the document under review. A network feed updated by the same system that wrote the manifest is not independent. A physical container tag, produced through a separate counting process, is independent. The value comes from that separation itself, not from the format or the guessed authorship of the document being checked.

Workers across many industries already keep independent references like this, even if they don't think of them as verification tools. A field technician's personal paper service manual has no live data link to the work order system. A pharmacist's printed drug reference sits outside the automated dispensing software. A structural inspector's stamped paper drawings predate the digital file's revision history. Each of those physical references does the tracing job that the automated system leaves undone.

The NIST Generative AI Profile published in 2024 identifies traceability as a critical governance requirement for AI-generated content, specifically because outputs frequently cannot be linked back to the source data used to generate them (Autio et al., 2024). Where a company's systems fail to provide that trace-back, the worker's independent physical reference fills the gap. A worker who reaches for the physical reference, instead of the matching digital record, is closing a verification gap the system itself left open.

Mapping High-Consequence Verification Targets

Work documents hold many data points, but most of them carry only small risk if they're wrong. A mislabeled storage zone code delays a single pallet. A formatting error on a routing slip costs one extra phone call to fix. Those errors are real, but a floor supervisor can catch and fix them inside the same shift. A different kind of error works under different stakes. Load specs carry consequences that a structural failure makes impossible to undo. A medication dosage error can reach a patient before anyone catches the mistake on a correction report. A wrong tax figure can create penalty exposure that a later correction doesn't automatically erase. The first practical task in manual verification is sorting those two kinds of error apart, before any one document even arrives.

High-consequence claims all share one key feature: a mistake in carrying them out produces a result that a later correction cannot undo. Physical consequences lock in the moment the work is done. A beam cut to the wrong dimension fails under load, and no updated work order changes what the structure already holds. Financial and legal consequences follow the same pattern, just on a longer timeline. A wrong tax figure creates a penalty that exists no matter what a corrected filing says later. A regulatory code applied the wrong way creates liability that an audit will find, no matter how polished or official the original document looked. These are the categories a verification map is built to target.

Building the map starts with one question: which errors, in my specific line of work, cause damage that a same-day fix can't undo? A worker in a physical trade points to claims tied to load tolerances and clearance specs. In clinical support, the priority targets are dosage entries and allergy flags. A financial worker fixes the map on tax rates and filing numbers. That short list becomes the target map. When a document arrives, the map tells the worker exactly where to look first, no matter who or what put the document together.

The carpenter working from an unlabeled cut-list shows how this targeting works at the task level. A standard cut-list holds dozens of entries: lengths, angles, quantities, material types, and assembly notes. Most of those entries can vary a little without causing a structural failure. A piece cut two millimeters short on a decorative trim run just wastes material. But a load-bearing beam cut to the wrong dimension sends the wrong force through the whole structure. The carpenter's verification map isolates the load specs first, because that category carries a consequence you can't undo. It doesn't matter that the cut-list has no label showing who made it; that doesn't change which category needs an independent check. Whether a senior estimator or an automated drafting tool prepared the sheet, the carpenter checks load specs against a stamped engineering drawing or a physical reference standard before the saw ever runs. How polished the document looks, and whether it carries a source label, has no bearing on that sequence.

The accountant working from an automated invoice payload works on a different timeline of consequence, but with the same map structure underneath. An invoice payload holds vendor information, line items, service descriptions, subtotals, and tax math. Most of those fields can be fixed through a revised invoice without creating any regulatory exposure. Tax figures work differently. Research on algorithmic systems has found that automated tools aim for strong statistical performance across large sets of data, not for guaranteed accuracy on any one output, which means a good overall accuracy rate can still hide individual, record-level errors (Kleinberg et al., 2018). For the accountant, that creates one specific risk. An automated invoice tool can turn out a payload with an overall accuracy rate that looks fine, while still getting one specific client's tax figure wrong. A tax rate applied the wrong way to a client's account creates penalty exposure that exists no matter what correction comes later. So the accountant's verification map isolates tax figures and filing numbers first, no matter which tool generated the payload, and checks them against the client's verified tax profile before the invoice goes through.

The mapping skill doesn't erase the work of verification. It focuses that work on the claims that carry the highest downstream cost, and it cuts the time cost of manual verification down to a level that pace pressure can absorb. A worker who has already picked out three or four high-consequence target categories in their field can scan an incoming document and reach those targets in seconds. That targeted scan does the same verification job a full-document review would do, without eating up the time that triggers a performance flag. The map turns source opacity from an unmanageable problem into a bounded, doable task.

Executing the Five-Step Cross-Check

The five-step cross-check turns manual verification from a loose habit into a set procedure. A loose habit bends under pressure. A set procedure holds its shape even when pace pressure, or noise and distraction on site, would cause a loose habit to skip a step. When a field technician is racing a service deadline in a loud, physically demanding place, a fixed sequence pays off because it doesn't depend on the technician remembering every detail in the moment. The sequence carries the structure, so the technician's full attention can go to the content.

Step one is claim isolation. The field technician gets a work order with a diagnostic code assigned to one specific piece of equipment. The work order also lists a job description, an assigned parts list, a time estimate, a customer contact name, and service history notes. Most of those fields just support office tracking; they don't drive the physical repair. The diagnostic code drives the repair. It names the fault the technician is being sent to fix, and it decides which repair steps the technician will follow. A wrong diagnostic code sends the technician down the wrong repair path on the wrong part of the system. Claim isolation means the technician picks out the diagnostic code as the one field that needs an independent check before any physical work starts, no matter who or what generated the work order.

Step two is source location. The technician finds the right section in the service manual before any comparison starts. Source location draws on a reference the technician already picked out ahead of time, not on a search done under time pressure. A technician who keeps a pre-built source map knows which manual covers which equipment family, and where its diagnostic code tables sit. The service manual carries one key property for verification: it has no live data link to the work order system. A digital service manual updated through the same software platform that generated the work order can carry the very same wrong code the work order does. The independent service manual predates that update, and it holds the original, manufacturer-verified code definitions. The technician opens the manual to the diagnostic code section before reading the work order data again.

Step three is entry comparison. The technician sets the work order in clear view and reads the diagnostic code straight against the code's definition in the service manual. This is not a general once-over. The technician reads the exact code, character by character, from the work order, and finds the exact matching entry in the manual's table. The manual's code definition describes the fault that code stands for. The technician then checks whether that fault matches the symptoms actually reported or seen at the service site. A diagnostic code generated by an automated work order system might carry a believable code number that actually points to a different fault than the one the equipment is showing. Entry comparison catches that mismatch, because the manual defines what the code means, no matter what the work order claims.

Step four is discrepancy logging. If the entry comparison turns up a mismatch between the work order code and the manual's definition, the technician logs it right away, before any repair work starts. The log entry holds four things: the clock time it was caught, the exact work order code as written, the manual's code definition that contradicts it, and a short note on the symptoms seen at the site. This log entry does two jobs. Right away, it gives the technician documented grounds to stop the wrong procedure and ask for a corrected work order. Over time, it builds an evidence record showing how often automated or upstream work orders produce diagnostic code errors across a service territory. The technician's discrepancy log is that feedback loop, running right at the point of work.

Step five is result recording. Whether the comparison turned up a match or a mismatch, the technician records the verification outcome before starting any physical work. A confirmed match gets its own note in the same log: the date, the work order identifier, the verified code, and the manual section that confirmed it. This result record closes the loop. If the repair goes ahead and a problem shows up later, the technician's documented verification proves that the technician independently checked the assigned code against the manufacturer's reference before any work began. That record is what separates a technician who followed a careful procedure from one who just accepted an unverified work order and did what it said. Those two positions carry very different consequences for a technician's career when a service failure comes under review.

With a marked reference and a prepared log sheet, the technician can run this whole sequence during the natural pause between arriving on site and starting physical work. That pause already exists in many field service jobs, right between stepping out of the truck and opening the equipment panel.

Calibrating Verification Speed Under Pace Pressure

Electronic performance monitoring systems track worker output in real time. A data entry clerk's keystrokes get measured against a target processing rate. A retail supervisor's floor setup time gets measured against an opening readiness standard. Supervisors, and the dashboards that watch them, get an alert whenever one worker's output drops below the threshold. Kalischko and Riedl found, in their 2021 review of electronic performance monitoring, that these systems create real, measurable worker stress, because the watching never stops. It removes the small bit of control workers used to have over natural pauses in their own workflow (Kalischko & Riedl, 2021). That loss creates a specific problem for manual verification. A worker who adds a visible pause to cross-check an incoming document risks setting off a throughput alert. The fix is to build verification into the task time the monitoring system already expects to see.

Many workflows have natural windows where the worker's hands pause between steps. Document loading, system lag, moving between workstations, and routine copying all create small gaps where a worker reads, repositions, or waits. Those gaps exist whether or not verification happens inside them. The monitoring system already counts their length as normal task time. A worker who uses these gaps to cross-check a high-consequence claim against a pre-positioned reference gets the verification done without adding any measurable time to the task record. The record shows nothing different from normal processing.

Pre-positioning is what makes that disappearance possible. A worker who has to search for a reference source while trying to verify something burns time outside the natural window. A worker who grabs a reference already sitting within arm's reach finishes the same check inside the window's normal length. Reference material placed at the workstation ahead of time, with the right sections marked and no page-turning needed, turns verification retrieval from an active search into a simple reach. That's the difference between a verification attempt that triggers a throughput alert and one the monitoring system never even notices.

The data entry clerk working an incoming batch file shows how this embedding works under constant electronic monitoring. The clerk's task is to open a source batch and enter its data fields into a core database. A reading step already exists inside the expected task time for every entry in the batch. The clerk uses that reading step to check batch totals against a pre-positioned reference ledger sitting beside the workstation. That ledger carries the prior verified totals for the account or file series the batch belongs to. When the clerk reads the incoming batch total, the eye moves from the source batch to the ledger line in one motion, then goes on to data entry, or flags a mismatch. The monitoring system records a keystroke pattern that matches what it expects to see. The verification happens inside the copy step itself, not after it as some separate task. If the batch total matches the ledger, the entry goes through. If it doesn't, the clerk marks the field, holds the batch back from processing, and logs the mismatch in a physical notebook before moving on. The clock shows no change from standard processing time.

The retail supervisor running floor setup before the store opens works under a different physical routine, but the same embedding logic applies. Floor setup means the supervisor moves through every store section and sets up displays at each station before opening. That same route already passes every station the supervisor would need to check for pricing accuracy anyway. So the supervisor builds the verification scan straight into the walking route, instead of treating it as a separate inspection. Before the setup walk starts, the supervisor prints a shelf price list and marks the high-consequence price points: promotional items, recently changed lines, and products flagged in the prior shift's exception log. At each station, the supervisor carries that printed list. While positioning signage or adjusting a display, that same stop lets the supervisor check the shelf tag against the printed reference. A price mismatch turns up during a station visit the setup walk already required. The supervisor marks it on the printed list and moves on. The setup task record shows normal floor preparation time. The verification scan created no separate time entry at all.

The clerk pre-positions the reference ledger before opening the batch queue. The supervisor prints and marks the price list before the setup walk starts. That prep work is what turns verification from a reactive pause into a built-in motion. Skip that prep step, and the worker has to improvise reference retrieval under pace pressure, eating up the window the monitoring system would otherwise count as standard processing time. This is how verification fits inside a pace-monitored job: the worker builds the verification structure before the task clock even starts, so the execution fits inside time the clock already counts.

Speed calibration comes down to doing things in the right order. A worker who identifies their natural task windows, pre-positions their reference material, and maps their high-consequence targets before the task even starts can run a full cross-check on a standard incoming document, all inside the time the monitoring system already allows for normal processing. The throughput record stays clean. The verification record closes. Both happen at once, because the prep work, done before monitoring even begins, erased the gap between them.

Exercises:

Exercise 1: Map Your Primary Verification Sources

Select the claim categories in your daily incoming documents that cause the most harm when acted on from wrong data. Record each category on a separate line in a notebook. For each entry, work through the following steps before you begin the task.

Identify one independent reference source that holds verified data for that claim category. A source counts as independent when it has no live data link to the system that makes your main work documents, and when no automated process can change its content without your direct say-so. A stamped paper drawing, a printed regulatory code table, a digital or bound manufacturer's manual, and a signed physical ledger all qualify. A digital file created from the same data that feeds your work order system does not.

Write the source's name and storage location beside the claim category in your notebook. Note the specific section or page range within the source that covers your claim target. Then record the maximum time it takes to retrieve the source and open it to that section from your normal work position. If that time exceeds thirty seconds, find a way to pre-position the source closer before the exercise ends.

Pre-position the two highest-priority reference sources at your workstation before you begin the task. Mark the relevant sections with a tab, clip, or bookmark so retrieval requires no page search. Test the retrieval: reach for each source without looking for it and open directly to the marked section. If the motion takes more than ten seconds, adjust the placement until it does not.

Review the map at the start of each week and update any entry when a reference source changes location, edition, or content.

Deliverable: A written, workstation-ready source map covering your top claim categories, with each reference pre-positioned and section-marked before you begin the task.

Exercise 2: Run a Timed Five-Step Cross-Check

Select one incoming document from your normal workflow that you have not yet acted on. A work order, an invoice, a shift log, or a routing manifest all serve this purpose. Set a timer before you begin.

Step 1: Identify the single highest-consequence claim in the document.

Step 2: Retrieve your pre-positioned independent reference source and open it to the relevant section.

Step 3: Compare the document's claim directly against the reference entry. Read both figures or descriptions aloud if the environment permits. Register whether they match.

Step 4: Open your notebook. If a discrepancy exists, record the document entry as written and the reference entry that contradicts it. Note the clock time of detection on the same line. If no discrepancy exists, record the document identifier and write "verified match" beside it.

Step 5: Record the full verification outcome before taking any action on the document. Close the notebook.

Stop the timer. Record the elapsed time at the end of the log entry. Repeat the full sequence on a second document in the same shift and record that time beside the first. The gap between the two times reflects retrieval and comparison efficiency rather than any flaw in the sequence itself. Most workers reduce their time on the second run once source retrieval becomes a physical habit. Run one timed cross-check per shift for two weeks and record each elapsed time in the same notebook column. Review the column at the end of week two and identify where time concentrates across the five steps. A step that consistently takes longer than the others identifies the point where additional pre-positioning or reference preparation will produce the most gain.

Deliverable: A two-week log of timed cross-checks with elapsed times recorded per shift, showing speed development and identifying the step that most benefits from further preparation.

Skill 2: Error Logging Procedures

The Architecture of the Error Log

The worker who catches an error in an incoming document, fixes it, and moves on carries the whole cost of that failure alone, with no trace left behind. An independent physical or digital error log breaks that invisibility. It does one thing: it turns hidden cleanup labor into hard, visible evidence. The log needs no software, no network access, and no vendor subscription; check your workplace's data and confidentiality policies before recording sensitive information in it. It just takes a physical notebook and a pen, plus the same set of columns used for every entry without exception. Workers can also build or use an independent digital log instead, as long as that data stays fully separate from the software and data that created the incoming document.

The structure of the log matters more than how it looks. Seven columns capture everything you need to build a record that holds up. The first column records the date and the exact clock time of detection. The second column records a document identifier, which can be a job number, a batch code, an invoice number, or any reference string that ties the entry to a specific incoming document. The third column records a plain-language description of the error, written in full sentences without abbreviation. The fourth column records the error category, classifying the type of failure as wrong figure, missing field, conflicting instruction, or corrupted data. The fifth column records the clock time the worker spent on cleanup labor, measured in whole minutes. The sixth column records any downstream impact the error caused before detection. The seventh column records the corrective action taken and the final resolution.

Those seven columns work well across most industries and job types, though some may need small adjustments. That uniformity is the point. Workers whose physical conditions, tools, and daily demands share almost nothing keep logs that share everything.

Consider the machinist. He receives a daily job sheet specifying torque values for a precision assembly. The figure is wrong. He catches it before the wrong number ever reaches the part, but only because he checks the figure against his own reference card. He opens his error log. He records the date and the time: 7:22 AM. He notes the job order number. He notes that the torque specification on the incoming job sheet conflicts with the manufacturer's published tolerance for that assembly. He classifies the error as a wrong figure. He records that he spent eleven minutes pulling his reference materials, verifying the correct value, and re-marking his work instructions. He notes that the error would have caused a failed torque test at final inspection if it had reached the assembly stage. He records that he corrected the figure manually before proceeding.

Eleven minutes doesn't sound like much on its own. But that machinist works forty-hour weeks, and if he logs two or three errors per shift across a whole month, that cleanup time adds up into a real, measurable labor cost. When workers fix upstream errors without writing down the time it took, that cleanup labor typically doesn't show up in the document system that caused the error in the first place. The error log is what brings it into view.

This consistency is what gives the log its power as evidence. A single shift's entry proves nothing on its own. A week of entries shows a pattern. A month of entries shows a system-wide failure rate. Real accountability needs documented human records that exist outside of the systems being checked. An independent error log, kept apart from any digital system the incoming documents touch, meets that requirement. As a physical document, it creates a record that no software update or dashboard change can erase or alter. It exists on paper, in ink, in the worker's own handwriting, with clock times no one else entered.

Categorizing Errors and Measuring Cleanup Labor

The four error categories in a standard error log exist for one reason: different kinds of error create different problems, and those problems carry different cleanup costs. A wrong figure sends a worker to a reference source to find the correct value. A missing field stops a process cold until someone finds or rebuilds the missing data. A conflicting instruction forces a worker to settle a contradiction before taking any physical action. Corrupted data needs a field-by-field comparison against an independent source before any record can move forward. Each category calls for a different response, eats up a different amount of cleanup time, and points to a different failure point in the pipeline that produced it.

This distinction matters, because sorting errors into categories without timestamps gives you incomplete evidence. A log entry that just says "error found, fixed it" tells no one anything useful. The same entry, with an error category, a detection time, and a clock-measured cleanup duration, tells management exactly which failure type ate up how many labor minutes, and how often that type shows up in a shift. That specificity is what turns one worker's story into real organizational data.

Timestamping takes discipline right at the moment you catch the error, not after the fix is done. When a worker catches an error, the first move is to write down the time before touching the document at all. Writing the time down after the fact adds guesswork that weakens the record. Writing it down at the moment of detection keeps an accurate starting point. The cleanup labor measurement then runs from that starting timestamp to the moment the corrected document goes back into the workflow. The gap between those two clock times, recorded in whole minutes, is the entry that counts as evidence. Rounding the time, or averaging it across a shift, weakens the record and makes it less useful as evidence.

The utility line worker shows what this looks like in a physical job with real safety stakes. His crew gets a daily field assignment sheet listing pipe location codes that show where excavation work will happen. A wrong location code puts machinery into the ground at a spot that doesn't match the actual pipe. The result can be a strike on a live utility line. Before the crew moves any equipment, he pulls the assignment sheet and checks the listed codes against the utility district's physical map binder, kept secure in the work truck. That binder holds printed, stamped records kept apart from the digital dispatch system that generated the assignment sheet.

He opens his error log the moment he spots the mismatch. He records the time: 6:51 AM. He classifies the error as wrong figure. Then he measures, in clock minutes, everything that follows: finding the correct code in the physical binder, marking the corrected value on the field sheet, telling the crew supervisor, and waiting for a verbal go-ahead before the equipment operator moves. That whole sequence takes nineteen minutes. He records nineteen minutes in column five. He records in column six that, without the correction, excavation would have started about forty feet west of the real pipe. He records the corrective action and the time the supervisor confirmed it.

Nineteen minutes of unplanned cleanup labor before 7:15 AM doesn't show up on the dispatch system's productivity dashboard. The dispatch system's own record shows only that the assignment went out and eventually got done. An uncorrected code, carried forward to tomorrow's assignment sheet, lets the same error come back with no flag at all. The log entry breaks that cycle by creating a dated, classified, clock-measured record that sits outside the digital pipeline making the errors.

A retail clerk fixing a wrong shelf price works under very different physical demands and hazards, but her log looks the same, because cleanup labor follows the same logic no matter the setting. Error type drives the response. The detection timestamp anchors the cleanup time. Clock-measured minutes turn one worker's labor into organizational evidence. Done the same way every shift, that pattern builds a record with a kind of weight that no dashboard summary can copy or erase.

Tracking Downstream Reach and Causation

An error in an incoming document doesn't stay where it lands. It travels. Every person who handles the document, and misses that specific error, becomes one more stop on the error's path through the operation. The log's sixth column, which records downstream impact, exists to map that path exactly: where the error started, how many process steps it crossed, what actions it nearly caused or did cause, and where a worker finally stopped it.

This mapping matters, because an error caught right at the point of entry means something different for the organization than the same error caught three or four steps later. An error caught early points to one single verification failure at the source. An error caught late points to a gap that runs across several process steps, each one letting the document move forward without an independent check. The log captures both realities, but only if column six gets treated with the same discipline as the error category and the cleanup time.

Tracking downstream reach means the worker has to piece together the document's path before writing the entry. That doesn't mean interviewing everyone who touched it. It means figuring out which process steps the document passed through, from its origin to the point where someone finally caught it. Each step marks a spot where one worker or one system passed the document along without independently checking that specific error.

The pharmacy technician at the dispensing window runs into this in a clinical setting, where downstream reach carries direct risk to a patient. A medication order enters the pharmacy system and moves through a set sequence: order entry, clinical verification, label generation, fill preparation, and dispensing. A staff member acts on the order's information at every one of those steps. When the label reaches the dispensing window with a wrong dosage figure, that figure has already passed through at least three earlier stages. The technician opens the error log right away. She records the detection time. She classifies the error as wrong figure. Then she rebuilds the path: the order entered the system, passed clinical verification, generated a printed label, and moved through fill preparation, before she caught the mismatch against the patient's physical medication record. She records in column six that the error crossed four process steps before detection, and that fill preparation had already started on the wrong dose before she flagged it. She records the corrective action, the supervising pharmacist's confirmation, and the time the corrected label entered the dispensing workflow.

That entry documents more than one error. It documents a four-step propagation gap, meaning four points in the process where an independent check should have caught a wrong figure and didn't. Bai, Nunez, and Kalagnanam, in their 2012 study, track how data errors move through a chain of business process tasks, and show that control points at each stage decide whether those errors keep traveling downstream. A dispensing technician who records that propagation depth in column six gives the organization real evidence of exactly where those rushed stages skipped a factual check. Without that entry, the incident record only shows that an error was caught and fixed at the window, with no view into how far it traveled before someone stopped it.

Entries across many industries can use the same basic column structure because downstream reach follows the same logic. An error travels as far as the system lets it.

Aggregating Shift Data into Organizational Findings

A single shift's worth of log entries answers one question: what went wrong today. A week's worth of entries answers a different question: what does the system keep failing to catch, and what does that failure cost in labor hours. Moving from daily entries to a weekly summary turns individual workers' personal records into organizational evidence that's harder to treat as isolated incidents.

The aggregation process can follow the same basic sequence across many industries. At the end of each shift, the supervisor collects the error logs from every worker on the floor or in the department. She counts the total entries across all logs for that shift. She totals the cleanup minutes from column five across every entry. She counts entries by error category from column four to find which type showed up most. She reviews column six to find which errors caused the highest downstream damage before detection. She records all four figures in a running weekly tally that builds up shift by shift until the weekly summary period closes.

That summary period produces a one-page document with five numbers on it: total errors logged for the week, total cleanup minutes, cleanup cost in dollars at the going hourly labor rate, the most common error category, and the worst downstream event recorded that week. The document's power comes from its numbers, not how it looks. One page, five figures, no extra story.

The manufacturing supervisor runs this process at the close of every Friday shift. His floor team works under a job documentation system that pushes updated spec sheets from a central software platform out to individual workstations all day long. The platform produces formatting that looks standard and complete. But it sends those figures to workstations with no independent check before delivery. His machinists log every mismatch they catch against their reference materials, recording cleanup minutes for each one. At the end of the week, he pulls every log from every workstation. He finds nineteen total entries across five shifts. Fourteen of those nineteen fall under the wrong figure category. He totals the cleanup minutes: four hours and twenty-two minutes of unplanned verification labor across the week. He works out the cost using the floor's going hourly rate, and records the total on the weekly summary sheet. He records the worst downstream event: one incorrect torque specification that had already reached the assembly stage before a machinist caught it during a reference check.

That summary sheet shows something the production dashboard doesn't. The software platform generates wrong figure errors at a rate of roughly three a day, and those errors cost the floor real labor hours that no output metric ever captures. Justesen and Plesner (2024) show how digital systems create hidden work: employees end up correcting bad data, cleaning up records, and bridging gaps the formal system leaves behind. That labor can stay almost invisible in how the organization talks about its own digital process. The weekly summary closes that gap by putting a real, verifiable dollar figure on cleanup labor that dashboards treat as if it doesn't exist.

In any workplace, the weekly summary does the same job. It takes five to seven minutes of tallying at the end of each shift, and turns that work into a running document that speaks a language many managers respond well to: labor costs in hours and dollars, tied to specific error types, with the downstream consequence recorded right beside each figure. Individual log entries build that document one shift at a time. The weekly summary is the point where the entries stop being one worker's personal protection and start becoming evidence the organization has to address.

Exercises:

Exercise 1: Build Your Worksheet and Make Your First Entry

Take any sheet of ruled paper, open a fresh page in a notebook, pull a blank page from a binder, or open a new word processing or spreadsheet file. Create seven vertical columns across the full width of the page. Label them: Date and Time, Document Identifier, Error Description, Error Category, Cleanup Minutes, Downstream Impact, and Corrective Action. Record your name and title or department at the top.

Select one document you handle regularly at work. Pick any figure, field, or instruction inside it that you would normally verify before acting on it. It does not need to contain an error. The goal is to practice the entry, not to manufacture a problem.

Record the current time in column one. Record whatever identifier appears on the document in column two. Describe the item you selected in column three using one full sentence. Assign an error category in column four. If no error exists, write zero in column five and "no error confirmed" in column six. In column seven, name the source you used to verify the figure, its date or version if one appears on it, and where you keep it or accessed it.

Confirm that every column contains an entry before you set the log aside. A blank column in any row is an incomplete record.

Exercise 2: Record the Summary Line

At the end of each week, record a single summary line directly below that week's last entry. If you encounter verification situations infrequently, record the summary line at the end of each month instead. Use whichever interval reflects how often you actually log entries; a weekly summary on a log with one entry every three weeks adds no value.

The summary line contains four items: the date range it covers, the total number of entries logged, the total cleanup minutes across all entries, and the error category that appeared most often. If every entry recorded zero minutes and no errors, record that. A clean summary is still a record.

Keep the summary line on the same page as the entries it covers. Over several weeks or months, those lines accumulate into a pattern you can read at a glance without re-reading every individual entry.

Skill 3: Localized Documentation Systems

Building the Offline Physical Record

The core skill of localized documentation is simple: you create and keep physical, offline, provenance-tracked records that live completely outside digital networks. These records protect institutional memory from two specific threats. The first is synthetic overwrite: automated content replacing accurate historical data with summaries a machine generated. The second is silent software rollout: system updates nobody announced, that alter, delete, or reformat existing records with no warning to the floor. A physical binder sitting on a shelf doesn't update overnight. A laminated field notebook sitting in a work truck doesn't sync with a central server at 2:00 a.m. The paper stays where you put it.

This matters because digital systems inside many companies are not fixed, unchanging archives. Software vendors push updates on their own schedule. Internal IT departments migrate databases, restructure file systems, and retire old storage formats. When those events happen, records can change in ways nobody on the floor ever approved. A digital job log from eighteen months ago might come back altered, cut short, or just gone after a system migration. A physical binder from that same period sits exactly as it was filed. The mechanism here is simple: physical storage sits outside every network event that can reach digital storage.

Silent rollouts create a specific hazard that many frontline workers run into from time to time but rarely put a name to. A procedure guide that read one way on Monday might read differently after a Tuesday update nobody announced on the floor. An automated scheduling system might quietly reclassify job codes in ways that shift accountability between departments. A digital service manual might drop a legacy equipment model from its database because the software vendor stopped supporting it. Workers who depend entirely on digital access have no independent baseline to check against. Workers who keep physical copies can still catch what changed, when it changed, and how far the change strayed from the prior, verified standard.

An agricultural operator managing field applications across several growing cycles runs into this problem in very physical terms. The operator records chemical application rates, soil amendment schedules, and irrigation timing in laminated field notebooks kept right at the site. Lamination matters for one specific reason: field notebooks face rain, mud, chemical splash, and swinging temperatures across a whole growing season. A standard paper notebook falls apart under those conditions within weeks. Laminated pages survive the full season without the ink fading or the pages falling apart. The operator writes entries in permanent marker, dates each one at the top of the column, and stores finished notebooks in a locked, weatherproof box at the field station.

Say a crop management software provider pushes an update that switches application units from gallons per acre to liters per hectare. The operator's physical log still holds the original units the team used out in the field. The software's conversion doesn't automatically fix old entries, and in some rollouts, it doesn't even flag the unit change for the end user. Without the physical notebook, the operator has no independent way to catch that mismatch during a regulatory review or a yield analysis. With it, the operator can lay the physical record next to the digital one. That shows exactly where the two split apart, and lets the operator document the cause with a dated entry that predates the software update.

Building the physical record takes two decisions up front, before the first document ever gets filed. The first is how you organize storage. Physical binders need a divider system that mirrors real operational categories, not calendar months. Sorting by equipment type, client site, crop field, or service route lets you pull things fast during a time-pressured service call or inspection. A binder sorted strictly by month gives a technician no advantage when they need specs for one specific unit in the middle of a repair. The second decision is what counts as eligible for the file. Workers set a clear intake rule for which documents get printed and filed: safety-critical specs, approved procedures, regulatory compliance records, and verified baseline data. Incoming unverified drafts, automated summaries, and system-generated reports sit in a separate staging area until someone confirms their accuracy against an independent source, before they ever enter the official physical record.

Implementing Provenance Tracking

Every document that enters a localized record system has an origin. That origin is one factor in how much independent verification it needs before a worker acts on it. Provenance tracking means putting a simple, consistent origin tag on each official entry the moment it's filed. The tag assigns one of two starting labels: Human-Observed or System-Generated. Human-Observed means a qualified person directly saw, measured, or confirmed the information through hands-on contact with the physical reality the document describes. System-Generated means the document came from an automated source: a software summary, an algorithmic output, or any digital process where no human directly confirmed the data before it went out. Those two labels sit at the top of every entry in the physical record, and they help guide how workers downstream handle that document.

The tagging system solves a real problem, one that gets worse the more automated tools a company layers onto its daily work. Workers reading paperwork can't tell from formatting alone whether a human or a software process produced a given entry. A system-generated patient summary and a nurse's handwritten shift note can both arrive printed on the exact same kind of paper. A software-built occupancy report and a front desk supervisor's direct room count can both land in the same inbox folder. Without an explicit origin tag, and absent other controls like signatures, review status, or audit trails, incoming documents can sit in the same unsorted pile, and workers either treat all of it as verified or invent their own personal habits for sorting it. Neither one produces a record that holds up. Provenance tracking replaces that guesswork with one formal rule, applied every time a document is filed.

A medical records clerk working in a small outpatient clinic handles incoming paperwork from several sources at once. Physician notes arrive dictated through voice recognition software and transcribed automatically, before any clinical review. Lab result summaries arrive as system-generated output from diagnostic equipment hooked into the facility's electronic health record platform. Nursing observations arrive as direct handwritten entries or chart notes made right at the bedside. Insurance correspondence arrives pre-filled by carrier software, formatted to look like someone reviewed it whether or not a human ever did. All of these documents land in the same patient file, and the clerk processes them at high volume, under time pressure.

When the clerk applies a provenance tag at filing, the whole process takes about five seconds per document. The clerk marks the origin in the upper right corner of the printed copy, using a two-letter stamp or a handwritten note: HO for Human-Observed, SG for System-Generated. A physician's dictated note, transcribed by voice recognition software with no clinical review, carries an SG tag until a physician co-signs it and the clerk updates the notation. A direct nursing observation, entered by hand at the bedside during patient contact, carries an HO tag. A lab result summary generated automatically by diagnostic equipment carries an SG tag, because the equipment processed the data with no human checking the output before it went into the record. The tag doesn't block the document from entering the file. It tells every worker who opens that file, later on, roughly how much independent confirmation the entry already has, and how much it still needs before clinical action can safely rely on it.

Provenance tagging needs one more decision built in: the update rule. When a System-Generated entry later gets human verification, the clerk adds a second note right next to the original SG tag. That second note records the name or employee number of the person who confirmed the data, how they confirmed it, and the date and time. That turns the entry from an unverified system output into a document with a verification trail anyone can trace. The original SG tag stays visible, so any auditor, supervisor, or regulator can reconstruct the document's full history, including its automated origin and the exact point where human review raised its status.

This two-tag system can work in many workplaces where documents come from a mix of automated and human sources, where local policy allows it. It doesn't need software integration or database access to start using, though it's worth checking with a supervisor before changing how documents are filed. A stamp, a pen, and a consistent filing rule are the only tools it takes. Its value shows up directly in the time it saves later workers, who would otherwise have to judge a document's credibility from scratch every time they open a file. A tagged physical record hands every later worker a clear starting point. It cuts down repeat verification work, and it shrinks the window where an unreviewed, system-generated error can travel unchecked through the operation.

Securing Tacit and Exception Knowledge

Every workplace holds knowledge that never makes it into any official document. A machine runs smoother if the operator lets it warm up three minutes longer than the manual says. A specific vendor always ships one unit short on orders over fifty pieces, and the receiving team just adjusts for it automatically. A particular billing code triggers an automatic rejection from one insurance carrier, unless a second modifier gets added, and only two people in the department know which modifier that is. This knowledge lives entirely inside the heads of experienced workers, passed along through watching each other work and informal instruction during shared shifts. Automated systems can't capture it, because they were never built to ask for it. When those systems replace or reassign the workers who carry it, the knowledge disappears right along with them.

Tacit knowledge is the term for skills a worker can't fully put into written rules. It builds up through repeated hands-on contact with a specific machine, process, client, or workflow, over months and years. Exception knowledge is the companion category: the built-up record of situations where the standard procedure fails, where the official documentation is wrong, and where a senior worker's workaround fills the gap. Both are institutional memory that no software summary can copy, because neither one was ever entered into any system in the first place. Securing them takes a deliberate, structured effort from the workers who hold that knowledge. That work has to happen before an automated process reassigns, relocates, or eliminates their role.

A plant mechanic on a manufacturing floor carries layer after layer of tacit and exception knowledge specific to each piece of equipment in their area. The official service manual for a hydraulic press covers standard maintenance intervals, approved lubricant types, and rated operating pressures. It doesn't mention that the pressure relief valve on unit seven starts to chatter at 2,200 PSI instead of the rated 2,400, or that running the unit at rated pressure without compensating throws off the cycle timing and knocks the downstream assembly sequence six seconds off pace. The mechanic learned this by watching it happen during a production run eighteen months earlier. No ticket ever got filed, because the unit never formally failed. The fix was an informal pressure adjustment that became standard practice for anyone assigned to that machine.

The mechanic writes this down in a dedicated section of the physical maintenance log kept at the workstation. The entry names the unit number, the behavior observed, the operating condition that triggers it, the fix, and the date the mechanic first confirmed it worked. The entry carries an HO tag, because it comes from direct, hands-on observation during actual operation. When the facility rolls out a predictive maintenance software platform that pulls sensor data and generates automated service schedules, that platform has no access to this informal adjustment history at all. It schedules the unit for service at standard intervals, based on rated specs. The physical log entry gives every mechanic assigned to the unit the real baseline the software can't provide, and it gives the maintenance supervisor documented proof of an equipment quirk the automated schedule will never catch on its own.

This documentation matters even more urgently whenever a company announces a new automated platform, a software upgrade, or layoffs in a department full of experienced workers. Those moments are exactly when tacit and exception knowledge faces the highest risk of being lost for good. A department that waits for the transition announcement to start documenting exception knowledge won't finish before the people who hold it are gone. The recording has to happen during normal operations, in real time, as the exceptions come up. That way the physical ledger builds up gradually across ordinary shifts, instead of in one rushed scramble after the company has already committed to a change it can't undo.

Establishing Data Storage Separation

The most direct way an unverified document contaminates an official record system is simple proximity, sitting too close to the real thing. When incoming drafts, automated payloads, and system-generated summaries share the same storage space as authenticated records, the line between verified and unverified information collapses. Workers pulling documents under time pressure grab whatever's closest and most recent. Supervisors reviewing files can't tell a confirmed entry from a draft that never cleared review. Auditors looking at the records find a mixed archive where origin and verification status are invisible unless they read every single document one by one. Data storage separation removes that risk. It builds a procedural wall that physically keeps incoming, unverified documents apart from authenticated official records, before any entry ever crosses into the official system.

That wall is a rule enforced through physical space, labeling, and an intake routinc. Incoming documents land in a designated staging area, whether that's a physical inbox tray, a labeled folder in a filing cabinet, or a separate binder marked clearly as unverified intake.

Where workers have the authority to do so, no document moves out of staging into the official record until someone completes an explicit verification step and applies the matching provenance note. The staging area works like a checkpoint, holding incoming material on purpose while the verification work happens. Once a document passes verification, it moves into the official record space. If it fails, or needs more confirmation, it just stays in staging until that gets resolved.

A legal secretary working in a law firm, a court support office, or a corporate legal department handles a high daily volume of incoming documents from many sources. Court filings arrive through electronic submission portals. Opposing counsel sends discovery materials through secure file transfer platforms. Automated contract review software generates clause summaries and redline comparisons. Client correspondence comes in through email and physical mail at the same time. Some of these documents carry real legal weight: a missed filing deadline, a wrong date on a court order, or an unverified clause in a contract summary can expose the firm, or its client, to direct legal liability.

The legal secretary sets up two physically separate storage spots for incoming documents. The first is a clearly labeled intake tray or folder for unverified material. Every document from any outside source, including court portals, opposing counsel platforms, and automated contract review tools, lands here first. The second spot is the official client file, kept in a locked cabinet or a secured, access-controlled digital folder that mirrors the same structure. A document moves from intake into the official client file only after the secretary runs a targeted review against the case record:

checking party names, dates, case numbers, and key figures against the court docket or the signed contract on file.

When an automated contract review platform generates a clause summary flagging a liability limitation clause, that summary lands in the intake staging folder. The secretary checks the summary's read on the clause against the original, executed contract stored in the official file. If the summary reads it correctly, the secretary moves it into the official file with an HO provenance tag reflecting that comparison. Automated contract review tools get clauses wrong at a measurable rate. When that happens, the secretary flags the mismatch, keeps the summary in staging with a note on the specific error, and alerts the supervising attorney before it can influence any legal work downstream. The wall between staging and the official file is exactly what makes that catch possible.

Exercises:

Exercise 1: Audit Your Department's Top Five Procedural Guides and Print Offline Copies

Identify the five procedural guides your department uses most often in daily work. These might include standard operating procedures, equipment maintenance checklists, regulatory compliance protocols, client handling instructions, or processing workflows. Don't pick documents just because they're official. Pick them based on how often your team actually reaches for them. The guides your team uses most during a shift carry the highest risk of disruption if a software update alters, moves, or removes them from the digital system with no warning.

For each of the five guides, complete the following sequence.

First, find the document in its current digital storage location. Record the file path, the platform name, and the date it was last modified, as shown in the system.

Second, open the document and check that the version stored matches the procedure your team actually follows day to day. If your team uses informal workarounds that stray from the written procedure, note those on a separate sheet attached to the printed copy.

Third, print a clean copy of the document at its current version.

Fourth, write the print date, the document version number if displayed, and your name or employee number in the upper right corner of the first page.

Fifth, file the printed copy in a physical binder for offline procedural references. Organize it by operational category, not by date.

After printing all five guides, examine the storage locations of each document and answer the following questions in your physical log.

How many of the five guides exist only in a digital format with no offline backup?

How many have been updated within the past ninety days without a formal announcement to your team?

How many reference equipment models, software platforms, or regulatory codes that your team no longer uses in practice?

Record your findings. These answers establish the baseline condition of your department's procedural documentation before the next unannounced software update changes it.

Exercise 2: Apply Provenance Origin Tags to Today's Shift Records

At the start of your next shift, designate a physical sheet or a section of your existing workstation log as the day's provenance tracking record. Create two columns. Label the first column "Document Description" and the second column "Origin Tag." You will use this sheet to classify every document you handle during the shift.

For each document you receive, process, or act on during the shift, record a brief description in the first column and assign one of two origin tags in the second column.

Write HO for Human-Observed if a qualified person directly saw, measured, or confirmed the information through hands-on contact, before the document reached you.

Write SG for System-Generated if the document came from an automated platform, a software summary, or an algorithmic output. Also use SG for any digital process where you cannot confirm a human checked the data before it was sent.

At the end of the shift, count the total number of documents processed.

Calculate the percentage of entries carrying an SG tag. Record that figure in your log with the date. For this exercise, treat a shift where more than half the documents carry an SG tag as a high-provenance-risk workload.

Review the SG-tagged entries and find any that led to a real action during the shift: a physical task, a financial transaction, a client commitment, or a regulatory record entry. For each of those actions, note whether you independently verified the document's key claims before acting on it. If you acted without checking first, record the specific claim you relied on, and the independent source you'd check next time. This last step turns the exercise from a simple classification task into a real planning document for your next shift.

Skill 4: Pre-Flight Checks

Identifying High-Risk Trigger Categories

Every incoming document carries many data points, but only a small number of them can hurt workers, destroy assets, or trigger legal trouble when they're wrong. The pre-flight check skill starts here: identifying those specific categories. A worker who knows exactly which figures cause the most failures can scan incoming paperwork in under sixty seconds, and catch the errors that matter most before acting on anything.

The term "high-risk trigger category" means any figure, measurement, instruction, or identifier that, if wrong, causes severe physical, financial, or legal harm downstream. These categories don't scatter randomly across a document. They cluster in predictable zones: numerical specs tied to physical tolerances, dosage instructions tied to medication thresholds, load ratings tied to equipment capacity, route parameters tied to infrastructure limits, and financial figures tied to regulatory filing rules. An automated document tool may generate a route plan, maintenance summary, or clinical chart note whose figures have not yet received an independent check.

This is the exact problem the pre-flight check addresses. Management mandates that speed up document throughput can reduce the time workers have available for verification. Say a fleet operations manager deploys an AI-assisted routing platform to cut planning time and dispatch labor. The daily route sheets drivers receive may look identical to sheets a human planner made. The formatting is consistent. The language is clean. The turn-by-turn directions read smoothly. But the clearance height assigned to a low overpass on a secondary route is a probabilistic guess, not a confirmed measurement pulled from a state infrastructure database. A driver running a vehicle with a loaded trailer height of fourteen feet, who gets a route sheet listing a clearance of fourteen feet six inches, cannot run that route safely without independently confirming that number. The sheet's polish offers no protection. Only the independent figure does.

A fleet driver working morning dispatch doesn't have unlimited time to cross-check every figure on a route sheet. The pre-flight skill doesn't ask for unlimited time. It asks the driver to know, before ever touching the paperwork, which specific category of figure carries deadly risk. For a driver running a heavy commercial vehicle, bridge clearance heights, posted weight limits, and permitted vehicle dimensions are the trigger categories. These are the three zones on any route document where one wrong number can end a run, destroy cargo, or put people in danger. Once the driver has mapped those categories, the scan goes straight to them, checks each one against an independent reference, and only then commits to the route. Everything else on the sheet gets a second, lighter pass.

Identifying trigger categories takes deliberate mapping, done ahead of time, away from the pressure of a live incoming document. Workers who build their category maps during downtime, or while onboarding into a new role, carry that map into every shift. When documents arrive at pace, the map is already active. The scan narrows straight to the categories that carry physical, financial, or legal weight. This is the foundation of the pre-flight check: know where the dangerous figures live before you even open the document, so when it arrives, you're already looking in the right place.

Designing the Sixty-Second Pre-Flight Checklist

The pocket inspection card is a physical tool. It sits at the workstation, clips to a clipboard, or folds into a shirt pocket. Its design comes from one simple principle: a worker under pace pressure can't reliably hold a mental list of high-risk targets while also processing incoming documents at the same time. The card puts that list outside the worker's head. It turns a mental task into a visual one. When a document arrives, the worker opens the card, runs each item against the paperwork in order, and either clears the document or flags it. When the card targets the right categories, the whole sequence takes under sixty seconds.

Building the card means choosing the right items. A checklist that tries to catch every possible error becomes a liability. It slows the scan, splits the worker's attention, and defeats the whole purpose.

The sixty-second pre-flight card holds no more than five to seven items. Each item names one specific high-risk category in plain language. Each item ties to a single action: check the figure, confirm the code, match the format, or compare the unit. Workers who write general instructions like "verify accuracy" on a pre-flight card end up with cards that don't work under real pace pressure. The card has to be specific enough that a worker can run it without stopping to think about what it means.

A public transit mechanic working a morning shift gets a daily maintenance sheet before touching any vehicle. That sheet lists the assigned inspections for each unit in service, including brake system checks, tire pressure readings, and any unresolved safety-related defects. The defect entries carry the highest risk on the sheet. A maintenance entry might show that corrective action was completed and the vehicle was cleared to return to service. But if a document generation system pulls that status from a prior shift's log, or matches it to the wrong vehicle ID, the sheet can show the defect as fixed while the vehicle still carries the actual problem. The transit mechanic's pre-flight card targets these entries first. The card item reads: "Safety-related defect entries: confirm each status matches the vehicle ID on this sheet and the corresponding maintenance record."

The card also carries items for brake pad thickness readings and tire pressure specs, because those numbers carry a physical tolerance where a wrong decimal creates an unsafe operating condition. The mechanic's five-item card covers the full trigger zone of the daily maintenance sheet. It doesn't cover every field on the document. It covers the fields where a wrong figure creates a consequence the mechanic can't undo once the vehicle leaves the bay.

The physical format of the card matters for reasons beyond convenience. When workers have to pull procedural steps from memory and carry them out under time pressure at the same time, their attention splits between remembering and doing. A physical card removes the remembering part entirely. The worker just reads the checklist, instead of rebuilding it from memory. In shift conditions where a new batch arrives every few minutes, or a maintenance sheet covers eight vehicles before a departure window closes, the card keeps the worker's full attention on the document itself, instead of on recalling the procedure for checking it.

Designing the card also means deciding on a format. The most useful pre-flight cards use a vertical list, with a checkbox or dash next to each item. Each item is written in plain language, fourteen words or fewer. The action to take is built right into the item, not listed separately. The card carries the worker's name, workstation, and the date it was last reviewed, so a supervisor can confirm it's current without breaking the scan cycle. Workers who update their cards whenever document formats change, or a new software rollout alters the fields they receive, keep the card useful through every system change a management mandate throws at them.

A pre-flight card built on the wrong categories, or drafted once and never touched again, gives false security. The card's value depends entirely on whether its items still match the current trigger zones in the documents the worker actually receives. That's why the card has to work as a living tool. Workers who treat it as a one-time thing lose its protection the moment the document format shifts. Workers who update it on the same schedule they use to check their physical verification references keep it useful through every rollout, every platform change, and every directive that lands with no warning from above.

Pattern Recognition for Probabilistic Errors

The pre-flight scan can't rely on formatting glitches or obvious inconsistencies to flag a suspect document. Workers train their attention to catch a different kind of signal: structural anomalies. These are internal features of a document that reveal something factually impossible, out of context, or numerically contradictory, even when the surrounding text reads perfectly clean. Structural anomalies don't announce themselves the way typos do. They take a trained eye to catch, and that training builds through repeated practice against the known error patterns in one specific work environment.

The most common structural anomalies fall into four groups. The first is internal numeric contradiction: a figure in one section of a document that conflicts with a related figure somewhere else in it. A total weight that doesn't match the sum of the listed line items, or a running count that resets wrong mid-page, is a sign that a generative system pulled figures from separate sources without cross-checking them against each other. The second is contextual mismatch: a spec, code, or identifier that belongs to a different record, time period, or asset than the one named at the top of the document. A part number pointing to a different equipment model than the unit listed on the work order, or a filing period code from last quarter showing up in a current-period report, means the system pulled data from an old record without confirming it still applied. The third is impossible precision, or rounding that doesn't make sense: a field that normally carries decimal detail showing up as a round number, or a field that's always round showing up with an unexplained decimal. The fourth is missing specificity: a field that should carry a named, counted, or coded value showing up with a vague, general term instead. A medication field that lists "analgesic" instead of a named drug at a specific dose, or an equipment field that lists "standard fastener" instead of a bolt size, is a sign the system filled the field with a plausible-sounding category label instead of a verified, specific value.

A catering operation getting a digitally generated event menu shows how these anomalies show up in a service setting. The document capitalizes course names correctly. Dish descriptions read smoothly. Quantities appear in the expected columns. A kitchen supervisor running a pre-flight scan doesn't read the menu for culinary quality; the scan targets structural anomalies in the operational data. The supervisor checks whether the portion counts scale correctly against the stated guest total. A menu listing forty portions of a plated entrée for a sixty-guest event has an internal numeric contradiction, right there. The supervisor checks whether any ingredient in a dish description conflicts with an allergen note in the column next to it, because a generative system building a dish description from a recipe database might pull an ingredient that triggers an allergen category the system, at the same time, marked as absent. The supervisor checks whether the prep times assigned to each course actually add up to a workable kitchen sequence, because a system generating each dish description on its own might assign overlapping prep windows that no single kitchen can pull off at the stated event time. None of these anomalies show up in the menu's formatting. All of them show up when a trained eye checks how the figures relate to each other, instead of just reading the text at face value.

Training pattern recognition for these anomaly types takes exposure to real examples from one specific work environment. Workers who collect flagged documents over several weeks, and review them for repeating patterns, build a working model of where probabilistic errors tend to show up in their document types. A technician who notices that torque figures in auto-generated inspection reports often fall outside tolerance in the rear suspension fields learns to scan that field first on every incoming report. A billing clerk who notices that account numbers in auto-generated invoice payloads often mismatch the client name by one digit learns to cross-check that pairing first on every incoming batch. This pattern library is specific to the work environment and the document types it produces. It builds through deliberate observation and direct experience, and workers who build it carry a pre-flight skill that no software update or management-directed platform rollout can take away, because the skill lives in the worker, not in the tool.

Integrating Pre-Flight Scans into High-Speed Workflows

Most workflows have intake moments: a brief, natural pause between receiving a document and taking the first physical action based on it. These pauses exist in most workflows, even under pace pressure. A warehouse order selector gets a pick-list before pulling the first item. An emergency dispatch clerk gets an incident summary before routing the first resource. The pre-flight scan doesn't add a new step to either workflow. It occupies the intake moment that's already there, turning a pause workers already used to orient themselves into a protected verification window. The difference between a scan that costs time and a scan that fits right inside the workflow comes down to one thing: whether the worker anchors it to the intake moment, or tries to squeeze it in mid-task.

Pace pressure in high-throughput jobs works through measurable performance numbers. Warehouse distribution centers track units picked per hour, route completion rates, and idle time between tasks. Emergency dispatch operations track call-to-dispatch times and routing accuracy. Automated management systems generate these numbers around the clock, and supervisors get dashboards that flag any deviation in real time. Workers who add an unscheduled pause mid-task can show up on those dashboards quickly, sometimes before they finish explaining why. A scan a worker completes before the task clock starts carries no penalty at all. A scan a worker inserts after the task begins reads as a deviation. The intake moment sits before that clock starts, and that's exactly what protects a worker's performance record while still getting the verification done.

A warehouse order selector working a fulfillment shift in a distribution center gets a pick-list at the start of each order cycle. The list arrives through a handheld scanner or a printed slip, depending on the facility's system. It tells the selector which items to pull, from which shelf location, in what quantity, and in what order. A warehouse management system generating these lists draws from inventory data that can carry stale location assignments, mismatched SKU codes, or wrong unit counts, if the upstream data feed had errors or a prior automated reconciliation created a conflict. The selector's pre-flight card targets three fields on every incoming list: the quantity for any item flagged as hazardous or weight-restricted, the SKU code for any item with a substitution restriction, and the shelf location code for the first pull in the sequence. The selector reads those three fields against the card, cross-checks the quantity and SKU against the physical shelf label at the first stop, and moves through the list. The whole pre-flight portion takes less than thirty seconds at the intake moment, before the selector even moves to the first location. That thirty-second window fits inside the natural pause every selector already takes to get oriented at the start of a new order cycle. It adds no measurable time to the task clock, and it catches the exact kind of list error that sends a selector to the wrong location, or gets the wrong item pulled into a shipment. The selector who checks trigger-category fields before moving doesn't slow the system down. The selector who pulls from an unchecked wrong location, and then has to reverse-route to fix it, costs the facility far more time than that thirty-second scan ever would.

Pace pressure turns dangerous when speed forces workers to act before the intake moment finishes. Management systems that squeeze document delivery and physical execution into one continuous motion eliminate the natural pause where verification happens. Workers who recognize the intake moment as a protected space, and anchor their pre-flight scan to it instead of treating it as some separate extra step, keep their verification capacity alive inside even the fastest workflows an employer can design. The pre-flight card makes that reality official: the scan belongs at the intake moment, because that moment exists in most workflows, regardless of the throughput target, and a worker who claims it owns the only verification window there is, before a wrong document drives a wrong physical action.

The pre-flight skill reaches its full value when workers combine four things: a current trigger-category map, a specific pocket card, a trained eye for patterns, and a fixed intake-moment anchor.

These four work together as one system. A current map makes sure the card targets the right fields. A trained eye catches the anomalies the card can't spell out by name. A fixed anchor makes sure the scan happens before the task clock starts, no matter the pace pressure coming down from above. Workers who hold all four carry a verification skill that survives most platform rollouts and shifting priorities around throughput and accuracy.

Exercises:

Exercise 1: Draft a 5-Point Pocket Pre-Flight Card

Pull three recent incoming documents from your workstation: any work orders, manifests, charts, batches, or instruction sheets you received and acted on within the last two weeks. Do not use documents you generated yourself. Review each document and write down the five fields or figures that, if wrong, would produce the most severe consequence in your specific work environment. Rank them by consequence severity. The field at the top of your list becomes item one on your card.

Write each item in plain language. Each item must name the specific field or figure and the verification action in a single sentence of fourteen words or fewer. "Torque spec: confirm against the manufacturer sheet in the bay cabinet" is a complete card item. "Check accuracy" is not. Work through all five items using the same format.

Format the card as a vertical list. Place a checkbox or dash before each item. Write your name, your workstation identifier, and today's date at the top. Print the card on cardstock or have it laminated. Clip it to your workstation, your clipboard, or carry it in your shirt pocket. If your document formats change at any point, update the card within the same week the change takes effect.

Exercise 2: Run a Timed 60-Second Scan on a Live Incoming Deliverable

Select a live incoming document at the start of your next shift before you take any action on it. Use a physical timer or the stopwatch on your phone. Set it for sixty seconds before you open the document.

Open the document. Run your pre-flight card against it in sequence. Check each of the five items on your card against the document's corresponding fields. Mark the checkbox when a field clears the check. If a field fails the check or produces a structural anomaly, circle the field directly on the document, write the anomaly type in the margin, and hold the document from execution until the discrepancy resolves through an independent reference.

Stop the timer when you finish all five checks. Record the elapsed time and the result in your error log: the document type, the date, whether all five fields cleared, and any anomaly you found. Run this exercise on at least one incoming document per shift, for two weeks straight. After two weeks, review your log and note which trigger category produced the most flags. If one category keeps flagging, add a second verification step for that item on your card. If the sixty-second window feels tight for a specific document type, find the intake moment in your workflow, and make sure you're starting the timer there, before the first task step begins, not after you've already committed to the document.

Skill 5: Handoff Verification Buffers

Mapping Natural Workflow Transitions

Most workflows already contain built-in pause points. These are moments where one person, crew, or department finishes a task and another begins. The framing crew drives the last nail and steps back before the crane operator lifts the beam. The departing nurse closes the chart and the incoming nurse picks it up. Work stops moving forward for a few seconds or a few minutes because the operational system demands it. No one installed these pauses to create extra work. They exist because physical safety, handover logistics, or scheduled shift cycles require them.

These pauses are the raw material of handoff verification buffers.

The problem is that most workplaces treat these moments as dead time. A framing crew finishes the rough structure and moves to a break area. The crane operator climbs into the cab, reviews the lift plan on a tablet, and proceeds. At no point does anyone ask whether the spec sheet the crane operator is reading matches the framing the crew just finished. Both parties worked from the same document source. That source may have been an automated plan that no human re-inspected after the last change order. If the document carries an error, both the framing crew and the crane operator absorbed it without a cross-check between them. The handoff passed silently, and the error traveled with it.

Silent handoffs are the main way unverified document errors travel through multi-stage workflows. When a document moves from one party to the next without a targeted confirmation question, every error inside that document gains a clean endorsement it never earned. The receiving party assumes the previous party already verified the content. The sending party assumes the receiving party will catch anything wrong. Neither assumption holds up, and automated document generation makes both of them dangerous. Polished automated output triggers the silent endorsement reflex, and the handoff completes without any real verification occurring.

Mapping natural workflow transitions solves this. It means finding every chokepoint where work already pauses, and treating each one as a pre-installed verification window. The mapping process requires nothing beyond a worker's existing knowledge of their daily sequence. The framing crew already knows when they finish their portion and when the crane operator takes over. The departing nurse already knows when the incoming nurse arrives and what documentation changes hands at that moment. The mapping process doesn't create the transition point. It just makes that point visible, as a place to ask one targeted question before the work moves forward.

The site framing crew and crane operator example carries real operational weight. A crane lift is a high-consequence, one-direction action. Once the beam goes up, correcting a placement error requires stopping the entire lift sequence, lowering the load, and re-rigging. In high-volume commercial construction, that sequence costs time measured in crew hours across multiple trade groups standing idle on site. The lift plan specifies load weight, pick point, swing radius, and set-down position. It often comes from a project management system that pulls structural specs automatically. Say the framing crew finished their section under revised dimensions, ones a change order introduced after the original automated plan was generated. The lift plan might still reflect the old numbers. The framing crew completed their work correctly. The crane operator received a plan that appears current. No one compared the two at the transition point.

A single targeted question at the handoff corrects that gap. Before the crane operator begins rigging, one crew lead confirms with a direct comparison: does the lift plan's set-down position match the framing crew's actual completed layout? That question requires roughly 90 seconds and two people. It catches dimension drift before a steel beam is in the air.

Mapping these transitions means a worker mentally walks through their daily sequence, and finds every point where they receive work from someone else, or pass work along. Each of those points is a natural chokepoint. The work pauses there because the process architecture requires it. The verification buffer does not add a new stop to the process; it converts an existing stop into an active confirmation step. That distinction matters, because it makes the buffer invisible to pace monitoring; some handoffs may still require checking with a supervisor before changing the process, since another worker is involved. The worker simply uses the pause that already exists to ask one question before the work continues.

Formulating Targeted Handoff Questions

The question itself is a precision instrument. A vague question at a handoff point produces a vague answer, and a vague answer provides no protection against an unverified document error. Workers who ask broad questions like "does everything look right?" give the receiving party no specific target to check. The receiving party scans the document visually, confirms the formatting is clean, and says yes. The handoff completes. The error inside the document completes the handoff with it. The question failed before anyone even answered it, because it asked for a general impression instead of a specific fact.

A targeted handoff question isolates one verifiable claim and asks whether that claim matches an independent physical reference. The question must be narrow enough that the answer is either yes or no. It must point to a physical source the receiving party can check within the time the handoff pause already allows. It must address the category of error most likely to carry operational consequence in that specific workflow. Workers in different physical environments, under different pace pressures, handling different document types, will write handoff questions that look different. That's because the high-consequence claims in their workflows are different too. The structural logic behind every targeted question is identical.

A freight receiver standing at a dock door receives an incoming shipment alongside a corresponding digital or printed manifest. The manifest specifies quantities for each line item. In a logistics environment, automated order management systems generate manifests from purchase order data. Those manifest figures may not reflect last-minute quantity adjustments made between the original order and the physical pick at the origin warehouse. The system generated the manifest from the original order. The warehouse picked from a revised allocation. The manifest did not update. The receiver accepts the shipment, signs the manifest, and enters the quantities into the warehouse management system. The error travels into inventory records and does not surface until a downstream pick order comes up short.

The targeted handoff question stops that sequence at the receiving door. The freight receiver asks: "Do these manifest quantities match the physical pallet tags?" That question points the receiving party to one physical reference: the pallet tags applied at the origin warehouse. It asks whether the figures on those tags agree with the figures on the manifest. The question produces a binary answer. If the tags match, the handoff proceeds. If they do not match, the discrepancy is visible before anything enters the system. The question takes under two minutes to answer, because it targets one specific category, quantity, and it needs a physical comparison, not a document re-read. The receiver does not review the entire manifest line by line from memory. The receiver holds the manifest against the pallet tags for the current shipment, checks whether the quantities printed on each tag match the quantities the manifest lists, and either signs or holds.

The single-item structure of the question matters beyond efficiency. A multi-item checklist at a handoff point splits the checker's attention across several claims at once, which raises the odds that one item only gets a partial scan. A single targeted question concentrates the full check on one high-consequence claim. If that claim passes, the handoff proceeds. If it fails, the entire handoff holds. The question functions as a gate, and the gate has one lock. That design keeps the buffer fast enough to survive pace pressure, while staying specific enough to catch the error most likely to cause downstream damage in that particular workflow. The question does not need to catch every possible error. It needs to catch the error most likely to cause the most damage if it crosses the boundary undetected.

Eliminating Silent Endorsements

A silent endorsement occurs when a worker receives a document and passes it forward without checking any of its content. The worker did not verify the document. The worker did not flag the document as unverified. The worker simply moved it to the next stage because moving it forward is the default behavior of the workflow. The receiving party at the next stage treats that forward movement as proof a check already happened. Nobody labeled the endorsement as silent. It became an endorsement the moment the document crossed the boundary without a question attached to it.

Silent endorsements are a product of workflow design. Many operational processes build speed into their structure, by assuming document accuracy was already confirmed at the point of creation. A kitchen floor runner does not write culinary orders. A records clerk does not draft the files they route. Both roles exist specifically to move documents from one point to another quickly. Speed is the operational measure of success for both positions. The expectation baked into each role is that someone else already confirmed the content before the document arrived. That expectation removes the verification question from the handoff before the handoff even begins. Automated document generation makes that assumption dangerous across the board.

When a probabilistic system generates a culinary order, the output carries the same visual markers as a human-prepared order. The dish names are spelled correctly. The quantities are formatted cleanly. The table assignment appears in the right field. None of those visual markers confirm whether the actual dish specifications, allergen flags, or modification notes are accurate. A floor runner who picks up that order and carries it to the kitchen does not evaluate dish specifications. The runner reads the table number, confirms the order is ready for delivery, and moves. That is the full scope of the role as the process defines it. If the order carries a wrong allergen flag, the runner passed it along with no idea a check was even needed. The silent endorsement completed the transmission.

Holding the order breaks that transmission before it produces a downstream consequence. A floor runner who pauses before delivering an unverified culinary order is not stopping the kitchen. The runner is using the seconds already built into the pick-up-and-carry sequence to ask one question: does this order reflect what the table actually requested? In a table service environment, the runner can check that question against the server who took the order verbally, or against a physical dupe ticket clipped at the pass. That comparison takes under a minute. If the order matches, delivery proceeds. If the table asked for a modification the automated system never captured, the runner catches it before the plate reaches the guest.

The physical act of holding is the operative skill. It takes a worker interrupting the automatic forward motion their role rewards, and inserting a check before picking that motion back up. In environments where speed metrics measure individual performance, interrupting forward motion carries professional risk. A runner who holds an order too long draws attention from floor management. A clerk who holds a file for review slows a routing queue. The buffer works precisely because the hold is brief and targeted. The hold lasts as long as it takes to answer one question against one reference, short enough to survive pace pressure if the question isolates the right claim in advance.

Structuring Handoff Verification Sign-Offs

A verbal confirmation at a handoff point disappears the moment both parties speak it. One party says the numbers match. The other says good. The work moves forward. Three hours later, when a downstream error surfaces, neither party can reconstruct exactly what they confirmed or who confirmed it. The exchange happened, but it left no record. A physical sign-off slip solves that problem by converting a spoken agreement into a written record that both parties produce, review, and retain at the moment of handoff.

The sign-off slip is a short, standardized form. The receiving party fills it out at the handoff point, and both parties sign it before the work crosses the boundary. The slip contains four elements:

the name of the deliverable changing hands, the specific claim the receiver verified, the name and signature of the person who completed the verification, and the timestamp of the exchange. Those four elements take under 90 seconds to complete. Together, they turn the handoff from an informal exchange into a documented transfer, one with a named verification owner and a clock-stamped record of exactly what the check covered.

Standardization is the structural feature that makes the slip operationally useful. A sign-off process that varies by shift, by supervisor, or by individual preference does not build an evidentiary record. It just builds a pile of inconsistent fragments, ones managers can't add up or use to make a real argument about document error rates or hidden cleanup labor. A standardized slip uses identical column headings and identical verification categories across every handoff point in a department. Every slip looks the same. Every slip contains the same information in the same position. That uniformity means any supervisor can read any slip from any shift and know exactly what the receiver checked, when the receiver checked it, and who owned the verification decision.

A warehouse dock runs on timed receiving windows. Carriers arrive within scheduled appointment blocks, dock doors open, and the receiving team has a fixed period to unload, count, and process each inbound load before the next carrier arrives. That pace creates constant pressure to accept and sign manifests quickly, which is exactly the condition that converts manifest errors into permanent inventory records. A 90-second sign-off slip inserted at the dock handoff interrupts that pressure at the moment the receiver signs the manifest.

The warehouse dock slip carries three fields. The first field records the manifest number and carrier name. The second field records the specific check the receiver completed: in this case, whether the manifest quantities matched the physical pallet tags for the current load. The third field records the receiver's name, signature, and clock time of the sign-off. The receiver completes the slip before signing the carrier's manifest. If the quantities matched, the receiver marks the verification confirmed and signs both documents. If the quantities did not match, the receiver marks the discrepancy and holds the manifest until a supervisor reviews the load. The slip becomes the physical record of which decision the receiver made and when. At the end of a receiving shift, the dock supervisor collects the slips and adds up any discrepancy flags into the shift log. That builds a running record of how often automated manifests carry quantity errors, on inbound loads from specific carriers or upstream systems.

Exercises:

Exercise 1: Design a 90-Second Physical Handover Sign-Off Slip

Take a single sheet of paper and draw four columns across the top. Label the first column "Deliverable." Label the second column "Verification Completed." Label the third column "Verified By / Signature." Label the fourth column "Time."

In the Deliverable column, write the name or identifier of the document, order, or file your role transfers at each handoff. Use the specific name your workplace assigns to that item, such as a manifest number, a work order ID, a patient room number, or a batch file name. Do not write a generic category. Write the actual identifier your team uses on that floor.

In the Verification Completed column, write one sentence describing the specific claim you checked before the handoff proceeded. Write that sentence in plain language a coworker from a different department could read without context. Do not write "checked document." Write "confirmed manifest quantities matched pallet tags for load #4782" or "confirmed dosage schedule in summary matched physician order on file for Room 14." The sentence must name the claim and name the reference you checked it against.

In the Verified By / Signature column, print your name and sign below it. If your workflow requires bilateral sign-off, add a second row below yours for the receiving party's name and signature before the deliverable moves forward.

In the Time column, record the clock time at which the verification completed, not the clock time at which the handoff began. The completion time is the operationally relevant record.

Once you have designed the slip, print ten copies. Use one copy at each handoff point during your next shift. At the end of that shift, review the completed slips and confirm that each Verification Completed field contains a specific sentence rather than a generic phrase. Any slip that carries a generic phrase identifies a handoff point where you need to sharpen the verification question before the next shift begins.

Exercise 2: Test a Single Verification Question at Your Next Shift Transition

Before your next shift begins, identify one handoff point in your daily sequence where work passes from you to another person, or from another person to you. Write down the single most consequential claim inside the document or deliverable that crosses that boundary. This is the claim whose error would produce the most downstream damage if it traveled forward undetected.

Write one question that asks whether that claim matches a specific independent reference. The question must be answerable with yes or no. The independent reference must exist physically or institutionally outside the document itself, such as a physical pallet tag, an approved purchase order, a physician's signed order, or a stamped delivery receipt. The reference must come from a different source chain than the document under review.

Bring the question to the handoff point at your next shift transition. Ask it out loud before the document moves forward. Record the answer on your sign-off slip. Note the clock time and the name of the party who provided the answer.

After the shift, note how long the question took to answer. If it caught a discrepancy, record the discrepancy in your error log with a timestamp and a description of the downstream damage the catch prevented. If it caught no discrepancy, the question still completed its function. A verified clean handoff is operational data. It confirms the document crossed the boundary with at least one high-consequence claim independently checked, which is a condition that unverified transmission cannot establish regardless of how clean the document appeared.

Run the same question at the same handoff point for five consecutive shifts. At the end of those five shifts, review your slips and logs together. Calculate how many handoffs carried a clean confirmation and how many carried a flagged discrepancy. That ratio tells you the error frequency at that specific boundary in your workflow. It also gives you the beginning of a data argument about what automated document generation costs your department in hidden verification labor per week.

Skill 6: The Format Bridge

Exploiting Required Translation Steps

Many workplaces generate mandatory translation steps. A carpenter transfers assigned measurements from a cut-list onto physical lumber before making any cut. A food service worker keys incoming digital orders onto a kitchen display board. A data processor copies field entries from a scanned form into a live database. These steps exist because raw incoming documents don't match the format workers actually use on the floor. Organizations require them to move information from one system into another, and automation rarely closes that gap completely. The translation step is a structural requirement, and it happens every shift regardless of who prepared the original document.

That structural requirement is also an opportunity. Every time a worker must physically transfer data from one format into another, they hold the incoming claim long enough to check it. The Format Bridge converts that required transfer into a dual-purpose window. The reformatting step doesn't add a separate audit task to an already packed schedule. The worker performs the transfer because the job requires it, and the verification happens inside the same motion.

Understanding why this matters requires understanding what happens when the translation step runs on autopilot. An incoming cut-list arrives on a job site. The document looks clean. The formatting is correct. The measurements appear in a clear, standard font with consistent decimal placement. A worker who copies those figures onto lumber without any independent check commits the whole job to those numbers, before physical reality gets a chance to prove them wrong. If automated scheduling software generated those numbers from an outdated project revision, the cut is wrong. The error does not surface at the point of document reception. It surfaces after the crew marks, cuts, and moves the lumber to the next stage of the build. Fixing it at that point costs far more than fixing it at the translation step.

This is where the carpenter example makes the mechanism visible. On a framing job, the carpenter receives a cut-list specifying board lengths for a structural wall section. The assigned task is to mark each board with a pencil line indicating where the cut must fall. That marking step requires physical contact with the lumber. The carpenter positions the tape measure at the board's end, reads the assigned measurement, and draws the line. The tape is already in hand. Checking it against the assigned figure from the cut-list adds no real time to the marking motion. The tape confirms whether the assigned measurement matches the physical layout on the framing plan pinned to the site wall.

The critical detail is that this check works regardless of who prepared the cut-list. Whether a project manager typed those figures by hand, or an automated estimating program generated them from a digital blueprint, the tape measure produces an independent physical result either way. If the cut-list says 92 and a half inches but the framing plan shows 94 inches in that bay, the worker sees the discrepancy before the cut happens. No separate audit layer appears on the schedule. The worker exploited a mandatory translation step already built into the job.

The conditions that make this skill necessary are easy to predict. When automated scheduling or estimating tools rely on digital blueprint files, errors already present in those files can carry forward into the generated measurements. A measurement that displays correctly on screen may differ from the stamped architectural drawing posted at the build site. A worker who treats the translation step as a passive transfer misses the only window where the error is still cheaply correctable.

The operational burden matters here. Carpenters on production framing crews work under tight board-count targets. Supervisors track cut output by the hour. A worker who stops to run a separate verification check, outside the normal marking routine, draws attention and risks a pace flag. The Format Bridge sidesteps that problem entirely. The tape is already in the worker's hand for the marking step. The framing plan is already pinned at the site. The dual-purpose window uses tools and actions that are already required, without adding any visible extra step to the workflow. The verification happens inside standard production time, and the marking output is exactly what a supervisor expects to see.

Catching an error at the point of physical transfer costs a fraction of what it takes to fix once fabrication or assembly has already begun. A miscut board on a framing crew can require ordering replacement stock, scheduling a second delivery, and rebuilding a wall section. The total labor and material cost of a single uncaught measurement error in that context can reach several hours of crew time. The tape check during the marking step takes under ten seconds per board. That ratio, prevention cost against correction cost, is what makes the Format Bridge a smart structural choice, not just a cautious one.

Structuring Dual-Purpose Reformatting Workflows

The Format Bridge becomes most reliable when the reformatting workflow itself carries the reference check inside its physical structure. A worker who receives incoming data, opens a reference ledger, and then enters the data into an operational system performs three separate steps. A worker whose reformatting layout puts the incoming data and the reference ledger in the same visual space during entry only performs two steps. The check happens because of the layout, not as some extra decision. Structuring the workflow so the reference is present during entry transforms a passive transfer into a built-in comparison.

A food service worker transferring incoming digital orders onto a kitchen display screen operates inside exactly this condition every time a shift starts. The incoming order arrives through a point-of-sale system or an online ordering platform. The worker's assigned task is to confirm the order details on the kitchen display so the line crew can read and execute them. In most kitchen environments, that display shows item names, quantities, modifiers, and special instructions. The worker inputs or confirms each field before the order reaches the line.

The reference material in a kitchen is specific and physical. Every location keeps a printed menu spec sheet, a laminated allergen chart, or a posted portion guide near the preparation area. These documents list every item the kitchen produces, the standard build for each item, and the ingredients that trigger common allergen concerns. The kitchen runs from these references because they're the verified record of what the operation actually produces, independent of whatever a digital ordering system reports on a given day.

The dual-purpose reformatting workflow places the incoming digital order and the kitchen's physical reference materials in the worker's direct line of sight at the same time. When the worker transfers a modifier from the incoming order onto the display, the posted spec sheet is visible in the same field of view. Say an automated ordering platform generates an order description listing an ingredient substitution the kitchen can't actually make. The worker sees that conflict during the transfer step, instead of discovering it when the line tries to build something impossible. A "no dairy" modifier on an item whose standard build requires a dairy-based sauce component does not survive the transfer intact if the spec sheet is present during entry.

A modifier conflict that survives the transfer reaches the line as a plausible instruction, and if an allergen drives the conflict, the guest safety exposure is active before any supervisor sees the error.

Structuring the dual-purpose workflow requires a deliberate layout decision made before the shift starts. The kitchen display transfer station needs the physical spec sheet and allergen chart within reach or in direct view. Workers should position reference materials so the physical document and the digital display occupy the same visual frame during entry. A worker who must physically turn away from the display to consult a binder on a separate shelf will skip the check under pace pressure. A worker whose reference materials sit right beside the display makes the comparison automatically, because both sources sit in the same field of view at the moment of entry.

The structural pressure in commercial kitchens creates a specific threat to this workflow. High-volume service periods pile up ticket stacks that push workers toward quick screen confirmation, without any real review. A kitchen running a hundred covers during a dinner rush operates under continuous speed pressure from the front of house, the expediter, and the ticket printer. Workers with no structured reference layout at the transfer station default to trusting the incoming digital order, because pulling a separate reference document during peak service costs real time right then. When the spec sheet already sits in the transfer station's sightline, the check requires no additional motion and imposes no pace penalty.

Peak volume periods create the conditions in which allergen and modifier errors become harder to catch before they reach the line. The dual-purpose layout addresses that vulnerability through physical structure. The worker does not need to decide to check the reference. The reference occupies the same space as the entry point, and the comparison happens during the transfer because the layout makes it unavoidable.

Embedded Cross-Checks During Ingestion

The ingestion step is the moment when incoming information moves from an unverified source document into an official operational record. Most organizations that run field operations, maintenance programs, or technical service work have an ingestion point. A field service technician copying assigned diagnostic codes from a dispatch ticket into an official work order executes that ingestion step multiple times per shift. The official work order is the record the organization commits to. It drives parts orders, billing, regulatory reporting, and subsequent service decisions. What enters the work order becomes the operational truth the system acts on going forward.

Template bridges transform that ingestion step into a validation gate. A template bridge is a structured entry form. It requires incoming claims to pass through specific field constraints before the system accepts them as official entries. The form may require a code to match a defined format, fall within a validated range, or correspond to an entry in a recognized lookup list. A field service tech entering a diagnostic code into a work order template, one that validates the code against a master equipment fault list, cannot accidentally import a malformed or outdated code without the template flagging the conflict. The ingestion step catches the error because the template's structure forces the comparison during the transfer.

To understand why this skill matters, it helps to understand how diagnostic codes reach the field in the first place. Modern service dispatch systems may receive fault codes or alerts from equipment telematics platforms and diagnostic scanners. Those codes identify conditions detected by the equipment or diagnostic system, but they do not replace the technician's physical inspection and diagnosis at the site. A code on the assignment ticket remains an input to the technician's work, not proof of the equipment's final condition. When a dispatch system pushes a diagnostic code to a field tech's assignment ticket, that code is just the automated system's best guess at what the sensor data means. It carries no guarantee it matches what the technician actually finds on site.

The field service tech's assigned task is to respond to the flagged fault, inspect the equipment, and document findings in the official work order. That documentation step requires copying the assigned diagnostic code from the dispatch ticket into the work order's fault code field. This is where the template bridge does its work. A work order template with a validated fault code field forces the tech to enter a code the system recognizes as current and correctly formatted. If the dispatch system generated a code that does not exist in the equipment manufacturer's official fault library, the template rejects the entry. The tech must resolve the discrepancy before the work order proceeds.

That resolution step is the embedded cross-check. The tech holds the incoming dispatch code against the validated list the template requires. That shows whether the assigned code is current, or carries a format error the automated system introduced. Then the tech makes a judgment about the equipment's actual condition before entering any code. This happens during the ingestion step, inside the time already allocated for work order completion. The tech does not add a separate audit task to the shift. The template's structural constraint makes the comparison mandatory as a condition of completing the required documentation.

The operational environment for field service work adds specific pressure to this mechanism. Field techs operate under response time targets. Dispatch systems track the interval between assignment receipt and work order submission. A tech who stops to resolve a code discrepancy outside the normal documentation process creates a visible gap in their response numbers. The template bridge addresses that pressure by embedding the resolution inside the work order entry step itself. The tech resolves the code conflict while filling out the form, and the form submission time reflects a standard documentation interval rather than a flagged delay.

The cost of uncaught diagnostic code errors in field service is direct and traceable. A tech who copies an outdated fault code into a work order without resolving the mismatch triggers a parts order against an obsolete repair procedure. Replacement components sourced to the wrong fault specification arrive at the job site incorrect. The tech must return for a second visit, the customer absorbs additional downtime, and the organization absorbs duplicate dispatch costs. Return visits caused by wrong initial documentation add costs a correctly documented first visit would have avoided entirely. The organization absorbs another dispatch, additional labor, and the consequences of extended customer downtime.

Exercises:

Exercise 1: Mapping Reformatting Windows in Your Daily Routine

Step 1: List Every Transfer Step You Perform in a Standard Shift.

Working from memory, or from a brief review of your daily task list, write down every step in your routine where you move information from one format into another. These steps may involve physical materials, digital systems, or both. Examples include copying figures from a printed order sheet onto a whiteboard, entering data from a paper form into a database, transferring measurements from a spec document onto a physical surface, or keying delivery information from a digital dispatch ticket into a work log. Write each step as a plain one-sentence description. Do not evaluate or rank them yet. The goal here is a complete inventory of every transfer action you do regularly, no matter how routine it feels.

Step 2: Identify the Source Document and the Destination Format for Each Step.

For each item on your list, write two additional pieces of information beside it: where the incoming information comes from, and where the information goes after you transfer it. The first entry is the source document or system. The second entry is the destination record, database field, physical surface, or display. This column pair clarifies the direction of each transfer and the operational stakes attached to it. A transfer that moves information into a permanent official record carries higher consequence than a transfer that moves information onto a temporary planning board.

Step 3: Identify the Reference Material Currently Available at Each Transfer Point.

For each transfer step, write down the physical or accessible reference material that exists in your work environment and could confirm the accuracy of the incoming claim during the transfer. This may be a printed spec sheet, a posted standard, a prior validated record, a physical measurement tool, or a manufacturer's reference document. If no reference material currently sits within reach or line of sight at that transfer point, write "none available" for that row.

Step 4: Mark Each Transfer Step as Active or Passive.

A transfer step is active if you currently compare the incoming claim against an independent reference during the transfer. A transfer step is passive if you copy or enter the incoming information without any comparison to an independent source during the transfer motion. Mark each item on your list with either "active" or "passive." Most workers find that the majority of their routine transfer steps are currently passive. The reason is rarely a lack of skill. The reference material is simply not positioned within the transfer window.

Step 5: Rank the Passive Steps by Downstream Consequence.

For each passive transfer step, write a brief answer to this question: what is the most immediate operational failure if an undetected error passes through this step? Rank your passive steps from highest consequence to lowest. The step at the top of your ranked list is your primary target for Exercise 2.

Exercise 2: Designing a Dual-Purpose Layout Sheet

Take the highest-consequence passive transfer step you identified in Exercise 1. This is the step where an undetected error carries the most immediate operational risk. The goal of this exercise is to design a physical or screen-based layout sheet that forces your highest-risk transfer step to function as a dual-purpose verification window going forward.

Step 1: Define the Transfer Fields.

Write down every specific data point you transfer during this step. Do not describe the transfer in general terms. List each individual field, figure, measurement, or identifier you copy from the source document into the destination format. A data entry clerk might list account numbers, billing period dates, and figure totals. A field technician might list fault codes, equipment serial numbers, and service intervals. A food service worker might list item modifiers, quantity counts, and allergen flags. The more specific this list is, the more precisely your layout sheet will catch errors right where they actually happen.

Step 2: Identify the Reference Source for Each Field.

For each field from Step 1, write down the specific physical or accessible reference document that carries the verified version of that data point. The reference must be independent of the incoming source document. A reference that draws from the same upstream system as the incoming draft does not provide an independent check. For each reference source, write its name, its physical location in your workspace, and whether it currently sits within your line of sight during the transfer.

Step 3: Draw the Layout.

On a blank sheet of paper, draw three columns. Label the left column "Incoming Claim," the center column "Reference Check," and the right column "Entry." In the left column, list each transfer field from Step 1 in the order you normally process them. In the center column, write the reference source for each corresponding field. In the right column, draw a blank line or box representing the destination entry point. This layout sheet becomes your visual frame during the transfer step. The incoming document sits to the left. The reference sits in the center. The entry point sits to the right. Your eye moves across all three during normal processing, and the comparison happens because of the layout, not as some separate decision.

Step 4: Run a Timed Test Transfer.

Using your completed layout sheet, perform one full transfer of a real incoming document at your workstation. Use a watch or phone timer to measure the total time from the first field read to the last field entry. Compare this time against your normal processing rate for the same transfer step without the layout sheet. For most workers, the difference per record runs under fifteen seconds. That difference is the total operational cost of the embedded cross-check. Write the result down and keep it.

Step 5: Resolve Any Reference Gaps Before the Next Shift.

If Step 2 produced any transfer fields with no available independent reference, address each gap before your next shift. This may require printing a reference document, requesting a validated prior-period extract from a supervisor, or physically repositioning an existing reference closer to your entry point. A transfer field with no independent reference remains a passive transfer regardless of how carefully the layout sheet positions everything else. The layout is only as strong as the reference material it places within the transfer window.

Skill 7: The Exception Flag System

Establishing Shared Team Error Notations

Most workplaces generate a category of incoming documents that should not move forward. A purchase order arrives with quantities that contradict the physical pallet sitting in front of the receiver. A maintenance instruction sheet lists torque values that a technician immediately recognizes as wrong for the equipment on hand. A clinical chart summary carries updated dosage numbers that do not match the original prescription on file. Workers who catch these problems all face the same risk: how to stop a suspect document from moving downstream, without taking personal blame for the delay. The hold must stay visible to colleagues, and the reason for it must stay clear.

This risk intensifies when workers cannot determine whether a document came from a human or an automated system. A fabrication shop may receive cut lists that production management software generated. A clinical unit may receive chart summaries that automated transcription tools populated. In both cases, the document arrives formatted and polished, with no visual indication that it skipped a human review step. The worker catching the error has no way to flag the source, only the problem itself. The system they use to hold that problem in place has to work no matter where the document came from.

The Exception Flag System converts private individual judgment into a shared, visible, team-level signal. The mechanism that makes this work is standardized visual notation: a physical marker attached directly to the flagged item or document. It sends the same meaning to everyone who sees it, no matter their shift, department, or role. Consistency is the primary design requirement for these markers. A marker signaling "suspect: hold for review" must carry that meaning every time it appears, without variation and without ambiguity. When a marker appears sometimes but not always, or when different workers apply different notations to signal different levels of concern, the system breaks. A colleague picking up a flagged item during a busy shift cannot stop to decode what a particular tag represents. The team has to agree on what the marker means, write that agreement down, and practice the system during low-pressure periods, before the first flag ever gets applied under real conditions.

In a manufacturing environment, the standardized notation typically takes the form of a high-visibility physical tag: a brightly colored card or hang-tag that attaches directly to the part or assembly. A machinist receives a cut sheet with dimensional tolerances that do not match the reference drawings in the offline binder at the workstation. The worker does not discard the sheet or set it on the bench where a passing colleague might mistake it for a completed item. The machinist attaches a high-visibility tag to the part, places it in the designated hold position, and that tag tells every person who passes the station that this item needs review before it moves forward. The part does not move, the cut sheet does not disappear, and the team sees the problem without anyone stopping to explain it verbally.

A visual signal at the point of detection stops an error from traveling downstream far more effectively than word of mouth alone. When workers catch errors but communicate them only by word of mouth, the signal degrades across shift changes, personnel rotations, and the pace-driven interruptions of a working floor. A physical tag attached to the item carries the signal forward without degradation, regardless of who stands at the station when the next step in the process begins.

Teams that skip the standardization step end up with improvised systems. Sticky notes fall off. Verbal descriptions lose detail by the third repetition. Items sit in ambiguous locations where their status stays unclear. Each improvised approach puts the verification burden back on individual memory, instead of spreading it across a shared, visible system. Standardized notation removes that memory burden from the equation and puts the signal directly on the item where every team member can read it.

The operational cost of a notation system is low once the team establishes it. A roll of brightly colored hang-tags costs less than a single rework cycle on a misprocessed order. A box of colored binder clips costs less than a single adverse event review triggered by an unverified chart update. The physical materials are cheap. The real investment is the time the team spends agreeing on what the marker means, and confirming every member understands the standard before anyone applies the first flag. That investment recovers itself quickly once the system operates as designed.

Constructing the Localized Exception Queue

A standardized visual marker stops a suspect document from moving forward. The marker alone, however, does not guarantee that the flagged item stays stopped. On a busy production floor or a fast-moving retail environment, a tagged item sitting on the wrong surface gets picked up, processed, and sent downstream by the next worker who assumes clearing the area is part of the job. The visual tag signals a problem, but a signal with no container fails once the shared work surface everyone pulls from becomes the default storage spot.

The localized exception queue solves this. It is a formally designated physical area, separate from the active workflow, where flagged items wait for qualified review. The queue is a specific location dedicated exclusively to items that a worker has flagged as suspect. Every item in that area carries an exception marker. No item in that area moves back into the active workflow without a qualified reviewer physically clearing it. The design principle is simple: physically separating the exception queue from the active work area makes it hard, not just discouraged, for a flagged item to move forward without authorization.

In a structural fabrication environment, this queue takes the form of an offline holding bay. A steel fabrication crew receives a batch of spec sheets through the job management system. A welder pulls a sheet for a load-bearing assembly, checks the material grade and dimensional tolerances against the reference binder at the station, and finds a discrepancy. The spec sheet calls for a steel grade that does not match the original engineering drawing. The welder attaches a high-visibility exception tag to the assembly, moves it off the active production rack, and places it in the designated holding bay. That bay might be a separate floor section marked with yellow boundary tape, or a physical cage rack the crew reserves for flagged work. That assembly does not ship. It does not advance to the next fabrication stage. It stays in the holding bay until a qualified structural reviewer examines the original spec against the tagged sheet and either authorizes the material or issues a corrected drawing.

The holding bay in this context functions as a physical firewall. The job management software that generated the bad spec sheet remains live and may continue pushing the same erroneous data to other stations. The bay does not fix the upstream software problem; it contains the downstream damage. Every assembly built from a suspect spec that a worker routes into the bay is a structural failure that never made it to a building site or a client installation. Catching a defect during fabrication costs a fraction of the remediation expense the same defect generates when inspectors or clients discover it after installation. The holding bay captures those costs at the cheapest possible point in the chain.

The maintenance requirements for a localized exception queue are minimal but non-negotiable. The holding area must carry clear physical boundary markers so that no worker accidentally stores active work inside it. It must carry a log, either a paper sheet mounted at the entrance or a physical index card system inside the bay, recording when each flagged item entered the queue, who flagged it, and what specific documentation issue triggered the flag.

That log turns the queue into a data source. When the same type of error shows up again and again across multiple items in the same queue over a week, the pattern in that log gives supervisors the evidence they need. It lets them escalate the problem from a local holding issue into a formal system correction request. The queue catches errors one item at a time, but the log reveals the structural failure producing them at volume.

Distributing Verification Capacity Across the Team

The exception queue holds flagged items in place. The notation system signals their status to every person who passes through the work area. Both tools solve the problem of visibility and containment. Neither tool resolves the flagged item. Resolution requires a reviewer with the domain knowledge to evaluate what the specific error means, where it likely originated, and what corrective action will actually clear it. When every flagged item goes to the same reviewer no matter its content, the review step becomes a bottleneck, full of errors the assigned reviewer may not even be qualified to judge.

Distributing verification capacity means pre-assigning routing paths, so complex flagged errors reach the right domain expert instead of just the nearest available supervisor. The routing assignment happens before errors surface, during the same low-pressure planning period when the team standardizes its visual notation and designates its holding areas. Each error category that the team defines as a trigger condition gets a corresponding routing destination: a named role, a specific person on each shift, or a designated escalation path. When a flagged item enters the queue, the worker who placed it there also marks the routing destination on the exception log. That way the right reviewer knows to check the queue, without waiting on a verbal handoff.

In a data operations environment, the routing gap between detection and resolution is often widest. A database manager oversees the daily ingestion of incoming records from an automated data pipeline. The pipeline pulls data from external sources and loads it into the organization's core system at scheduled intervals. Some records arrive with corrupted fields: account numbers that fail format validation, date entries that fall outside acceptable ranges, or transaction figures that contradict the established data pattern for the corresponding account. The database manager applies the exception flag, and routes the corrupted records to the designated holding queue, separating them from the verified batch before the ingestion process finishes. The detection step falls within the database manager's scope; the resolution step does not. Figuring out whether the corruption comes from a source-side formatting error, a misaligned field mapping in the pipeline configuration, or a logic fault in the ingestion script takes the technical depth a senior data analyst carries. The database manager routes the flagged records to the senior analyst, and includes the exception log entry noting the specific field errors and the batch timestamp. The senior analyst then examines the records against the pipeline configuration to find the actual source of the problem. The database manager does not attempt a fix. The senior analyst does not need to search the floor for the flagged batch. The routing path connects detection to the right resolution without losing time or accuracy in the transfer.

This routing structure also protects both workers individually. The database manager who flags and routes a corrupted batch creates a documented record of the detection. The senior analyst who reviews and resolves the batch creates a documented record of the resolution. Neither worker carries unilateral responsibility for the full error cycle. Errors that a reviewer without the right domain expertise tries to resolve tend to come back, because the fix addresses the visible symptom, not the actual cause. When automated pipeline errors repeat, the accumulated log entries across both roles give the team a documented pattern that justifies a formal correction request to the system administrator or the software vendor.

Utilizing Formal Organizational Exception Channels

Individual flag events resolve one item at a time. A tagged item enters the holding bay. A routed record reaches the domain expert who can clear it. Each resolution handles a single document and returns one item to the active workflow. The local flag system manages these events efficiently when they occur in isolation. When the same type of error surfaces across multiple items, multiple workers, and multiple shifts, the local system has reached its limit. The source of the problem sits upstream, in the automated or organizational system that generated the flawed documents. The local flag system has no way to reach it. Formal organizational exception channels fill this gap.

Many organizations keep override protocols, deviation reporting forms, or system ticket mechanisms on hand; check whether yours does. These exist for situations where standard procedures can't handle an anomaly. Field workers typically encounter these channels in the context of safety incidents or regulatory compliance events. The same channels apply when recurring documentation errors from automated systems cross the threshold from isolated mistakes into a pattern of systemic failures. Turning a pile of individual flag events into a formal system ticket places the problem into the organization's formal review process, instead of just absorbing the cost one document at a time.

The exception log that workers maintain throughout the flag process becomes the primary evidence base for formal reporting. A log entry records the date, the specific document error, the time required to detect and contain it, and the downstream risk it carried before the worker flagged it. When the same field appears corrupted across a series of log entries over a two-week span, those entries constitute a documented pattern. The formal reporting system requires exactly this: a description of the anomaly, the dates it occurred, its operational impact, and the corrective actions taken at the local level. Workers who have maintained the physical log from the beginning can file a complete, evidence-backed report. Workers who resolved individual errors without logging them have nothing to bring to the formal channel beyond a verbal account.

A utility field technician provides a direct example of how this escalation path works in practice. The technician receives daily work orders from an automated dispatch system that assigns equipment inspection routes using location codes tied to the utility's infrastructure database. Across several shifts, the technician has flagged multiple work orders whose location codes do not match the physical infrastructure maps stored in the service truck's paper binder. Each flagged order went through the standard local process: exception tag applied, order held, technician verified the correct location against the physical map before proceeding. The exception log now carries six entries across nine shifts, each recording a different location code error on a different work order, all tracing back to the same dispatch system. The technician compiles these log entries and files a formal deviation report through the utility's quality management system. The deviation report references each log entry by date and order number, describes the specific field error in each case, and records the total verification time the errors required across the reporting period. The quality management system generates an official ticket, assigned to the dispatch system's technical team. They have to investigate the location code data feed and document what they find within a set resolution window.

This feedback mechanism is what the local flag system cannot provide on its own. An automated dispatch system pushing wrong location codes has no way to know that field technicians have been correcting its errors by hand for two weeks, unless someone files a formal report. The system continues generating the same output because no organizational signal has reached the team maintaining it. Formal exception channels create that signal. They turn a pile of individual corrections into real institutional pressure for an upstream fix. And they create a documented record that protects every worker who added a log entry along the way.

Exercises:

Exercise 1: Define Three Mandatory Exception Flag Triggers

Most workers carry an informal sense of which document details tend to cause problems. They know from experience that certain figures routinely turn out wrong. They know certain fields get corrupted during data transfers. They know some instructions arrive with no supporting reference to make them workable. The first exercise converts that informal sense into a written, standardized list of three specific conditions. Each one will trigger a mandatory exception flag every time it appears in incoming documentation, with no judgment call required at the moment of reading.

Start with the documents you handle most often during a standard shift. Pick the work orders, chart summaries, manifests, batch reports, or instruction sheets that arrive at your workstation with enough regularity that you can recall their structure without looking. For each document type, write down the fields or data points that carry the highest operational consequence if they contain an error. These are the targets: the figures that, if wrong, cause rework, safety exposure, financial loss, or downstream failures before anyone catches the mistake.

From those targets, define three specific, observable conditions that will each trigger a mandatory flag. Precision is the standard. A condition must name a specific field, a specific type of discrepancy, and the action it requires. A phrase like "figures that look off" fails this standard, because it needs judgment at the moment of reading instead of a testable comparison. Write each condition as a single sentence. Examples of the required specificity: "Any incoming torque specification that falls outside the range listed on the reference drawing at this station triggers a mandatory flag." Or: "Any medication dosage figure on an updated chart summary that does not match the standing order in the paper nursing record triggers a mandatory flag." Or: "Any quantity field on an incoming manifest that differs from the physical count on the corresponding pallet tag by more than two units triggers a mandatory flag." Each sentence names a field, describes the discrepancy, and specifies the response.

Transfer these three conditions onto a pocket card. Index cards work well. Laminated paper works better if the workstation environment involves moisture, oil, or heavy handling. Post the card at the station where incoming documents first arrive in your workflow. The card must be visible without searching for it. Run the next three incoming documents through the three conditions before those documents move forward in the workflow. Note how long each scan takes. If a single scan runs longer than thirty seconds, revise the condition description until it points to one specific field rather than requiring a broad read of the full document. The card's value comes from speed and precision, not from comprehensiveness.

Exercise 2: Set Up a Physical or Digital Holding Queue

The second exercise requires building the holding infrastructure that keeps flagged items in place once an exception trigger fires. This queue has to exist as a designated location, separate from the active workflow surface, before the first flag event ever happens. Teams that improvise a holding area after the first flag event create the exact confusion and risk the exception flag system is meant to prevent: items end up in ambiguous locations, their flagged status disappears with shift changes, and the exception log has no fixed address to point to.

For a physical holding area, identify a specific location in your immediate work area that does not share surface space with active workflow items, checking with a supervisor first if you don't have the authority to designate the space yourself. A separate shelf, a labeled physical inbox tray, a marked floor section with boundary tape, or a dedicated rack position all serve this function. The location must be reachable from the primary workstation without leaving the immediate area, and it must be visible to any colleague who enters the space. Attach a paper log sheet to the front of the holding area before the area receives its first flagged item. The log sheet needs five columns: the date, the document or item identifier, the specific flag condition that triggered the exception, the name of the worker who applied the flag, and the name or role of the reviewer the item requires. Fill in the column headers now, before the queue is active.

For a digital holding area, create a dedicated folder or queue location in whatever document management system your workstation uses, if your role allows you to create one; otherwise, ask an administrator to set one up. Name it something unambiguous, like "Pending Review" or "Exception Hold." If you have administrative access, set the folder's permissions so designated reviewers can open the contents without requesting access during a time-sensitive event; otherwise, ask your system administrator to set this up.

Create a matching digital log with the same five-column structure: date, item identifier, flag condition, flagging worker, assigned reviewer. Pin the log to the folder, so anyone who opens the holding folder sees the log without a separate search step.

Once the physical or digital queue is in place, run a single test cycle before the next live shift begins. Pull a document from your most recent batch of incoming work and apply the full exception flag sequence: identify the trigger condition using the pocket card from Exercise 1, mark the document with the exception notation your team uses, move the document to the holding area, and complete the log entry in full. Time the full sequence from trigger identification to log completion. Most workers complete this in under two minutes once the queue exists in a fixed, accessible location. If the sequence takes longer, find which step causes the delay, and adjust the physical layout or the log structure until the time reaches something sustainable. The holding queue functions as a live operational tool only when applying the flag and completing the log entry take less time than the downstream cleanup work the queue prevents.

Skill 8: Collaborative Audits

Mapping Institutional Expertise Across the Floor

The idea behind Collaborative Audits starts with a simple fact: no single worker carries complete domain knowledge across every part of a complex work document. A job packet for a commercial renovation might contain electrical load calculations, structural framing specs, and equipment anchoring requirements on the same two pages. A clinical handoff summary might compress medication adjustments, wound care instructions, and dietary restriction updates into a single printed sheet. When those documents arrive with clean formatting and institutional authority, they move fast. Workers trust them because they look finished. The danger sits precisely in that polish.

Nothing on the surface of the page shows whether a credentialed professional reviewed the figures, or automated software just assembled them. Under that condition, the only reliable defense is human domain expertise applied in a targeted, structured way. Collaborative Audits formalize that defense by turning floor-level expertise into a verification resource before a document commits to physical execution.

The first step in building that defense is mapping the expertise that already exists in the room. Most operational floors carry institutional knowledge distributed unevenly across individual workers. Formally credentialed professionals carry one layer of it. Workers who have accumulated years of hands-on repetition carry another layer that no certificate documents. Both layers carry verification value that no automated summary can replicate.

On a commercial construction site, that expertise distribution becomes visible the moment work begins. The master electrician who has wired dozens of industrial panel rooms carries load calculation knowledge that goes well beyond the published code tables. He knows which conductor gauges run hot under real field conditions, which breaker configurations the local inspector flags regardless of what the code technically permits, and which spec-sheet shorthand conceals assumptions that only hold in controlled lab settings. When an incoming electrical plan arrives, his review of the load figures and conduit routing takes thirty seconds of focused reading. That thirty-second check covers a verification gap that would take an untrained worker hours to find using reference materials alone.

Mapping this expertise requires a deliberate walk through the floor's actual human resources. The process answers three questions for every role present. First, what specific document categories does this person have the hands-on experience to verify?

Second, what are the most common error types that document category produces under automated generation or rapid drafting conditions? Third, how quickly can this person run a targeted check on a flagged section without pulling them off productive work for more than a brief interval?

That third question carries real weight, because collaborative verification fails whenever it disrupts the floor instead of blending into it. A floor supervisor who routes every flagged document to a single senior specialist transforms that specialist into a bottleneck. The goal of mapping is to distribute verification capacity across multiple domain experts so that complex documents route to the right reader for each specific component. An incoming multi-trade maintenance packet might send its electrical figures to the master electrician, its structural details to the lead carpenter, and its calibration notes to the senior equipment tech. Each reviewer examines only the section within their domain. Each check runs in parallel. The document clears faster, and the verification covers more ground than any individual worker could manage alone.

The mapping process also surfaces which document categories the floor currently cannot verify internally. That gap carries direct operational significance. A team that knows it lacks a domain expert for a specific document type can take deliberate action: it can hold the document until a qualified reviewer is available, escalate through official channels, or route the flagged section to the formal quality management process. A team that hasn't mapped its expertise gaps just processes the document by default, and discovers the error later, downstream, under full operational pressure.

Physical trades carry one version of this problem. Clinical environments carry another. Logistics operations, regulatory compliance offices, and food service production floors each carry their own variation. The expertise distribution looks different in every setting, but the structural logic of the gap stays consistent across all of them. Complex documents contain embedded claims that require domain-specific knowledge to verify. That knowledge exists in the workforce. The workforce has to know where it sits before it can deploy it.

Documenting the map for your own reference requires no formal org chart revision or management-approved skills inventory; using it to formally route work to other reviewers may still need a supervisor's sign-off. A shift supervisor with a pocket notebook and twenty minutes can walk the floor, ask three questions per worker, and produce a working verification reference covering the team's actual expert resources. That reference belongs in the same offline binder as the error log and the pre-flight checklist. It is a tactical tool.

The master electrician on the construction site is not an exceptional case. Most operational floors carry equivalents. Workers with that depth of hands-on experience represent verification assets that go entirely unused when no one has mapped where that expertise sits. The first task of Collaborative Audits is to make it visible, locate it precisely, and keep it ready before the next complex document arrives.

Structuring Micro-Consultations

A micro-consultation operates on a fundamentally different scale from a formal peer review. It does not require a conference room, a scheduled block on a shared calendar, or a formal agenda. It just takes one worker walking a few steps to reach a peer with the right domain knowledge, asking one precise question about a specific claim in an incoming document, and recording the result before heading back to work. The entire exchange takes thirty seconds when the question is properly formed. That constraint is structural, not aspirational. The thirty-second limit forces the worker to isolate the exact claim that requires expert confirmation before approaching the peer. That isolation step is where the verification discipline actually begins.

On a commercial kitchen floor, that routing happens constantly under conditions that leave no margin for extended review. The lead line cook occupies a position where incoming documentation directly controls what goes onto a plate. An allergen sheet for a modified menu item or a catering event arrival list identifying guest dietary restrictions carries immediate physical consequence. Say the sheet wrongly lists an ingredient as allergen-free, because an automated generation tool drew from a standard recipe template and missed a procurement substitution. That error travels straight to the guest. The lead line cook cannot verify every ingredient claim against physical stock during active service. What she can do is isolate the claims that carry the highest risk, identify the one person whose domain knowledge covers that risk category, and ask one precise question before committing the dish to production.

The question she asks does not request general feedback on the document. She holds the allergen sheet, points to the specific substituted ingredient listing, and asks: "Does this match what came in on Tuesday's delivery?" The kitchen manager who signed off on the substitution order or the prep cook who handled the delivery physically knows the answer. That exchange takes fifteen seconds. The lead line cook records the confirmation in her shift log and proceeds. If the answer contradicts the sheet, she flags the document before any preparation begins. The micro-consultation converts a credible but unverified claim into a confirmed data point without pulling either worker off the active line for more than half a minute.

Keeping the exchange at or below thirty seconds also protects both workers' performance metrics. A shift supervisor who sees a lead line cook and a kitchen manager exchange ten words over a sheet of paper registers the interaction as routine kitchen coordination. A compliance clerk who stops at a colleague's desk for twenty-five seconds reads as normal floor communication. Neither worker flags as unproductive. The verification happens inside the ordinary flow of work, so it reads as normal activity rather than an unauthorized interruption.

The structural design of a micro-consultation requires precise preparation. The worker isolates one claim and identifies the domain expert who covers it. The question must be specific enough to produce a confirmable answer. When any of those pieces is missing, the exchange drifts past its time boundary and stops functioning as a daily verification tool. Preparing before the approach matters just as much as the confirmation itself, because a poorly framed question wastes the domain expert's time and teaches both people that the process costs more than it's worth.

Building Cross-Training Verification Networks

A micro-consultation routes one claim to one expert. That exchange resolves a targeted uncertainty quickly and returns both workers to their immediate tasks. A cross-training verification network operates at a different level of document complexity. It handles incoming work orders that span several technical domains at once, where no single expert reviewer can cover the full scope of embedded claims without blowing past the time limits that make peer verification workable day to day.

Work orders that span several domains create this problem hardest. An automated repair instruction packet for industrial equipment might combine parts specifications, diagnostic step sequences, and calibration tolerances on a single document. The parts specifications require someone who knows the current inventory, supplier substitutions, and cross-reference codes. The diagnostic sequences require someone who has performed the actual repair procedure under field conditions. No single worker covers both domains with the depth required to catch probabilistic errors in each one. A cross-training verification network assigns each domain to the right reviewer and coordinates the review so the document clears without becoming a scheduling problem.

The field service tech and parts specialist pairing shows how that coordination works in practice. A field service tech who services and repairs industrial HVAC units carries deep procedural knowledge about repair sequences. He knows which diagnostic steps produce unreliable readings on aging equipment, which calibration procedures require field adjustments, and which repair sequences require judgment because actual installations do not always match the conditions assumed in the written procedure. What the field service tech does not carry with the same depth is current parts inventory knowledge, supplier catalog accuracy, or the substitution codes that apply when a specified part is backordered against an acceptable alternative.

The parts specialist carries exactly that inventory depth. She tracks which manufacturer part numbers have changed in the current catalog cycle, which aftermarket substitutions meet OEM specifications under the service agreement terms, and which part codes an automated system is most likely to pull from an outdated template. When an incoming repair instruction packet arrives with a parts list attached, she reads the codes, cross-references them against the live inventory system and the current supplier catalog, and finds a substitution error or a discontinued part number in under two minutes. She cannot verify whether the diagnostic step sequence in the same packet applies correctly to the specific unit generation listed on the work order. That verification belongs to the field service tech.

The cross-audit splits the packet between them along those domain boundaries. The field service tech takes the procedural pages. The parts specialist takes the parts list. Each reviewer reads only the section that falls within their area of expertise. Each brings a targeted question to the review. The field service tech asks whether the diagnostic sequence matches the unit generation specified and whether the calibration tolerances reflect field installation conditions. The parts specialist asks whether the listed part numbers match the current catalog and whether any substitution codes apply to the active inventory. They compare findings before the packet clears to execution, note any discrepancies in the shift log, and route confirmed errors to the formal exception channel.

The structural requirement for this pairing to work is deliberate design before the document arrives. A cross-training verification network does not form when a complex work order appears on the desk. It forms in advance, during a quiet period, when a supervisor or lead worker maps the document categories that regularly arrive against the domain expertise available on the team. The mapping identifies natural pairings based on complementary knowledge gaps. A field service tech and a parts specialist cover one document type. A licensed electrician and a structural lead on a renovation project cover another. A senior clinical pharmacist and a unit charge nurse cover a clinical protocol update. Each pairing addresses a specific multi-domain document pattern that the team processes on a recurring basis.

Building the network also requires both paired workers to understand what the other reviewer covers and what they do not. The field service tech needs to know which sections of the repair packet fall outside his verification scope, so he routes those sections to the parts specialist instead of passing judgment on claims outside his domain. The parts specialist needs enough functional understanding of diagnostic sequencing to recognize when a procedural claim contradicts the parts specifications she is reviewing. That cross-literacy does not require full expertise in the other's domain. It requires enough working familiarity to recognize when a cross-domain conflict exists and flag it for joint review before either worker signs off.

The joint sign-off is the measurable output of the network. A repair instruction packet that clears through a single reviewer carries one verification point of authority. The same packet cleared through a paired cross-audit carries two independent domain confirmations with two logged names and two timestamps. When an error escapes one reviewer's domain knowledge, the second reviewer's log entry documents which section each person reviewed and which claims each person confirmed. That record protects both workers from individual accountability for errors that fell outside their assigned review scope. Spreading verification labor across two specialists at once cuts individual metric risk and increases the total coverage depth the team gets on complex documentation.

The cross-training network also builds operational resilience over time. The field service tech who regularly cross-audits with the parts specialist develops enough inventory literacy to recognize obvious part code anomalies during his own initial scan. The parts specialist develops enough procedural familiarity to flag a diagnostic sequence that looks structurally wrong before the packet even reaches the field tech's desk. That accumulated cross-domain awareness does not replace the formal pairing. It accelerates the review because both workers arrive at the cross-audit with pre-screened documents rather than unread ones. The network sharpens the people inside it while it closes the verification gaps that automated document generation consistently leaves open.

Diffusing Individual Metric Risk Through Peer Review

A performance system built around documents processed per hour, tasks completed per shift, or records entered per session measures one thing: throughput. If accuracy checks and downstream correction costs never show up in that same metric, a worker who pauses to verify looks less productive than one who just keeps processing without stopping. That asymmetry creates direct employment pressure against verification behavior, regardless of what the organization's stated quality policy says about accuracy.

Automated document generation can intensify that pressure, because expectations can shift toward machine-speed processing.

The solution is structural reframing. A worker who pauses alone to verify a document creates an individual performance gap that the monitoring system flags. Two workers who pause together to coordinate on the same document create a team coordination event that the same system reads as operational communication. The verification activity does not disappear. The individual metric risk distributes across the pair, and the visible output shifts from a solo slowdown to a recognized workflow exchange. The goal is to turn an individual verification pause into a documented team coordination event. Peer review gives that pause a visible purpose, instead of leaving it as unexplained lost output.

On a physical trade crew, that conversion happens naturally within the communication patterns already present on the floor. A lead carpenter on a commercial framing crew who spots a dimensional anomaly in an incoming cut-list does not need to leave the floor, consult a manual alone, or quietly set the document aside. He turns to the senior framer standing two feet away, holds the sheet between them, and asks one confirmation question about the load spec in question. That exchange takes twenty seconds. A supervisor walking the floor sees two experienced workers discussing job documentation before cutting begins. The activity registers as professional due diligence. The lead carpenter's daily output log reflects the verification time as crew coordination, which it functionally is.

Digital processing teams face a more granular monitoring challenge because activity logs, keystroke metrics, and queue timestamps track individual output at a finer resolution than a floor supervisor's visual observation. A data entry clerk who pauses between records to run a verification check produces a gap in her keystroke log that the system can identify and flag. The reframing principle stays the same, but making it work takes deliberate structural alignment. When the clerk routes a flagged record to a senior data analyst for a targeted review, the routing event generates a system timestamp. The analyst's confirmation generates a second timestamp. The combined log entry documents a two-person quality review rather than a solo processing pause. When the queue system records exception handling as a normal workflow event, the routing and confirmation timestamps place the verification pause inside the documented work process, instead of leaving it as an unexplained gap in individual activity.

The joint sign-off is the operational anchor that makes this reframing durable. A verbal confirmation between two workers leaves no record. A written or digital sign-off creates a documented coordination event with names and timestamps attached. On the physical trade crew, the joint sign-off is a two-initials entry on the job sheet before cutting begins. On the digital processing team, it is a dual-confirmation field in the queue management system or a timestamped annotation in the record log. Both outputs turn the verification pause from an invisible individual behavior into a visible team activity, one the organization's own documentation systems can recognize and defend.

The overall effect of consistent peer review shows up in shift-level error summaries. If fifteen minutes of peer-review time prevents forty minutes of downstream correction labor, the net productivity calculation favors the review. The cleanup labor that individual-speed processing generates piles up in hourly blocks that rarely show up in the same performance report as the throughput figures that caused it. A shift supervisor who tracks both numbers sees the relationship clearly. That data belongs in the error log alongside the individual verification records, where it becomes a quantitative argument for the team's coordination activity.

The workers who benefit most from this reframing work under the tightest surveillance: piece-rate processors in back-office environments, and many workers whose output logs or queue times feed automated performance dashboards. For many of them, the individual verification pause carries real employment risk when monitoring systems treat accuracy work as lost production time. Turning that pause into a documented peer coordination event gives verification an organizational form the monitoring system can read as legitimate work, because within any functioning quality framework, it is exactly that.

Exercises:

Exercise 1: Build Your Peer Domain Expert Inventory

Pull out a pocket notebook or a blank page from your offline verification binder. Walk your floor, workstation area, or office during a low-activity period, such as a shift break, a pre-shift window, or the first quiet minutes of a slow production cycle.

For each colleague present, record three data points. Write the person's name and role in the left column. In the center column, list the specific document categories that person handles with enough daily repetition to verify claims independently. List document types rather than general competencies: repair instruction packets, allergen sheets, contract clauses, parts manifests, maintenance logs, or whatever categories your workplace processes regularly. In the right column, write the most common error type that document category produces. If you do not know the error pattern yet, leave that field blank and fill it in after the next error log entry involving that document type.

Review the completed inventory for coverage gaps. Circle any document category your team regularly receives that no single colleague can verify with direct domain expertise. Those gaps require escalation to management or a formal hold protocol when a document in that category arrives with suspicious figures.

Store the inventory as a single laminated reference card or a tabbed page in your offline binder. Review and update it at the start of each month or immediately after any team staffing change. An inventory that reflects last quarter's team composition carries no operational value for this week's document load.

Exercise 2: Execute a Timed Peer Audit on a Complex Spec Sheet

Select one incoming document from your current workload that contains claims spanning more than one technical domain. A repair instruction packet with both a parts list and a procedural sequence qualifies. A contract with both payment terms and regulatory reporting obligations qualifies. A clinical summary covering both medication adjustments and dietary restrictions qualifies. The document must be one you would normally process without a second reviewer.

Read through the document and mark every claim that falls outside your direct domain expertise. Use a pencil mark in the margin. Do not evaluate those claims at this stage. Mark them and move on.

Identify the one peer reviewer from your expert inventory whose domain expertise covers the highest-risk marked claim. Form a single question in writing before you approach that person. The question must name the specific claim, identify the specific section it appears in, and reference the independent data point you want confirmed. Write it out in full. A question you cannot write in one sentence is not specific enough to produce a confirmable answer in thirty seconds.

Start a timer. Approach the reviewer, present the marked section, and ask your written question. Record the answer and stop the timer when the exchange ends. Log the reviewer's name, the timestamp, the specific claim reviewed, and the confirmed or flagged result in your shift log. If the exchange runs past sixty seconds, the question was too broad. Revise the question format before running the next audit.

Run this exercise on one complex document per shift for five consecutive shifts. At the end of the five shifts, review the log entries. Count how many claims required a flagged follow-up against how many received a clean confirmation. That ratio establishes your baseline for collaborative audit frequency on that document category and gives you quantitative grounds to present peer review time as documented quality coordination during any performance review.

Skill 9: Informed Refusal Protocols

The Legal and Professional Foundations of Informed Refusal

The ninth skill in this system means understanding something most frontline workers already sense by instinct, but rarely know how to carry out with real legal protection: that in certain conditions, they have the right to stop. Informed refusal is the documented, formal act of withholding a high-risk consequential action when incoming paperwork contains specific, verifiable errors that threaten safety, financial assets, or legal compliance. Workers who carry out a bad order, just because it arrived with formatting polish and an authoritative tone, bear the physical and legal consequences of that action. The skill defines the operational boundary that keeps those consequences where they belong.

That boundary carries real legal weight. In the United States, the Occupational Safety and Health Act of 1970 grants workers the explicit right to refuse work they reasonably believe presents imminent danger. The key word is "reasonably." A worker cannot simply feel uncomfortable with a task. The legal standard requires that a reasonable person, knowing the same facts, would conclude the danger is real and immediate. This standard shifts everything from subjective discomfort to documented evidence. That means the skill of informed refusal can't be separated from the skill of evidence collection. Workers who log errors in a bound ledger, timestamp discrepancies, and preserve original documents carry legal standing. Workers who walk off a job with only a gut feeling carry almost none.

Professional nursing ethics operate on a similar logic. The American Nurses Association's Code of Ethics makes clear that technology supports nursing practice but does not replace the nurse's knowledge, skill, or responsibility to the patient: a nurse carries a direct duty to the patient, and that duty does not transfer to the system that generated a bad instruction. When a medication order arrives on a chart summary and the dosage figure fails a basic clinical cross-check, the nurse's duty to verify before administering is not optional. Whether a physician typed the chart summary or a clinical documentation tool generated it, the verification requirement stays the same. The nurse's professional responsibility and the patient's safety depend on that verification happening before the medication reaches the patient.

The structural pressure that makes informed refusal difficult is the speed mandate. Facilities, warehouses, construction sites, and clinical floors often operate under staffing ratios and throughput targets that treat verification pauses as delays. Supervisors under throughput pressure pass that pressure downward. Workers absorb the expectation that faster means better because stopping feels like falling behind. That expectation is operationally wrong. Incorrect, unverified, or incomplete work instructions injure workers. A speed mandate that suppresses verification is, itself, an injury mechanism.

The excavation worker makes the structural mechanics of informed refusal visible. An excavation crew receives a trenching sheet specifying soil classification, excavation depth, and required shoring type. That sheet may arrive as a printed PDF from the project office or as a handwritten note from the foreman. The worker on the ground has no way to tell, just from format or presentation, whether a licensed geotechnical engineer verified the soil classification, whether someone copied it wrong from a prior project, or whether an automated scheduling tool generated it by carrying a previous job's numbers forward without adjustment. A trenching failure in Class C soil treated as Class B soil does not give a worker time to respond. Cave-in collapse happens in seconds. The excavation worker's grounds to raise a refusal grow strongest the moment the soil classification on the incoming sheet cannot be verified against an independent, physical soil test on record, though the formal legal right to refuse depends on additional OSHA conditions being met. That refusal is a documented hold pending confirmation of a single, high-consequence data point, and OSHA's excavation standards under 29 CFR 1926.652 provide the regulatory foundation that makes that hold enforceable.

Some form of legal and professional infrastructure for this skill already exists across many regulated industries, though the specific protections vary. Workers do not need to build it from scratch. They need to know it exists, find the specific formal channels their industry uses, and understand that using those channels counts as professional compliance, not workplace resistance. The documentation habit, the channel knowledge, and the willingness to use both form the operational foundation on which every other component of this skill depends.

Distinguishing Resistance from Quality Management

The most common reason a worker's refusal fails comes down to framing. Workers who present their concern as personal opinion risk having it read as a behavioral issue rather than a quality issue. Workers who present their concern as documented evidence give management a systemic quality failure to correct. The difference between those two readings determines whether a worker ends a shift with a written warning or with a formal corrective action in the system that protects both the worker and the organization from downstream liability.

Framing a refusal correctly means stripping out anything that sounds like attitude, and anchoring every statement to a specific, verifiable fact. That is the structural requirement for a refusal to function as quality management. A statement like "I don't feel comfortable with this route" gives a dispatcher nothing to work with and gives a supervisor every reason to treat the concern as a performance issue. A statement like "This route directs the vehicle through a tunnel on Route 9, and 49 CFR 397.67 requires applicable hazardous-material routes to avoid tunnels unless no practicable alternative exists" closes the conversation differently. One is a position built on feeling. The other is a documented regulatory conflict with a code citation attached.

This distinction matters, because organizations process these two kinds of input through entirely different channels. An emotional position is more likely to route to HR. A documented regulatory conflict routes to safety management, legal, or compliance. The routing channel determines who reviews the refusal, what authority they carry, and whether the organization treats the underlying document error as an individual behavior problem or a systemic operational failure. Workers who understand this channel logic use it deliberately. They frame every refusal so that it lands in the channel best equipped to evaluate it on the merits.

The hazmat driver provides a precise example of how this framing works in daily practice. A driver hauling Class 3 flammable liquids or Class 1 explosives operates under a layered regulatory environment. The U.S. Department of Transportation's Pipeline and Hazardous Materials Safety Administration, known as PHMSA, requires carriers to pre-plan and document routes for certain hazardous materials shipments under HM-232 rulemaking standards. Title 49 CFR Part 397 sets specific driving restrictions for vehicles transporting hazardous materials by highway, covering tunnels, bridges with posted weight limits, densely populated areas, and alternative routing requirements around certain infrastructure. The route sheet a driver receives at the start of a shift must comply with those regulations before the vehicle moves.

Automated routing software changes that compliance picture. Commercial routing platforms calculate paths using cost and time variables. Some platforms include HAZMAT restriction filters, but those filters require correct configuration at the administrative level. Say a dispatcher uses a platform with an improperly configured HAZMAT restriction layer, or a platform update resets a restriction setting back to default. The system then generates routes that look valid but violate federal transportation law.

At that point, the driver's refusal to execute that route is the operational intervention that prevents a federal violation, a potential accident in a restricted zone, and the carrier liability that follows both. The driver refuses because the route sheet directs the vehicle to a point that federal highway code prohibits. That refusal requires four things to function as quality management: a specific identification of the non-compliant route segment, a citation of the applicable regulation by number, written notification to the dispatcher and shift supervisor through the company's official safety reporting channel, and a documented request for a corrected and compliant route sheet before departure. None of those steps involve personal opinion. Each one anchors the refusal to verifiable data.

The organizational effect of that framing is measurable. A refusal grounded in a 49 CFR citation forces the carrier's safety officer into the loop. It creates a paper trail showing the driver identified a regulatory conflict, reported it through the correct channel, and withheld action pending correction. PHMSA enforcement records show that carriers face civil penalties for hazardous materials routing violations, with fines for individual incidents reaching into the tens of thousands of dollars. A driver who documents a routing conflict before the vehicle moves protects the organization from that exposure. A driver who runs the non-compliant route, just because stopping the dispatch felt like resistance, absorbs the operational and legal consequence of the organization's system failure.

Quality management framing converts a worker's pause into institutional protection. It forces the correction upstream, where the route sheet originated, before the consequences land downstream. It keeps the driver's conduct record clean while generating organizational evidence of a systemic document failure. Workers across regulated industries carry this same leverage whenever they hold a verifiable regulatory citation against a non-compliant incoming document. That leverage only activates when the worker uses the formal channel, documents the specific conflict, and anchors the refusal to the evidence.

Executing the Pre-Action Refusal Sequence

The pre-action refusal sequence is a specific, ordered procedure. It does not begin with a conversation. It does not begin with a supervisor's approval. It begins with documentation, because documentation is the foundation that makes every subsequent step defensible. A worker who skips straight to notification without a written record of the specific concern relies on memory and goodwill to protect their professional conduct. A worker who documents first creates an evidentiary baseline that the organization cannot easily rewrite after the fact.

The sequence runs in three steps. The first step is documenting the concern with enough specificity to make the error actionable. That means recording the exact field or data point that carries the error, the value the incoming document shows, the value that an independent source confirms it should show, and the source reference that establishes the correct value. General statements like "the numbers look wrong" do not constitute documentation. Statements like "Employee ID 4471 shows 96 hours worked in the pay period ending 03/14; confirmed timecard shows 80 hours for the same period" do. The distinction is not stylistic. A documented error with a specific data point and a source reference triggers a procedural response. A vague concern triggers a supervisory judgment call, and that call frequently favors the schedule.

The second step is notifying supervision through the organization's official reporting channel. For most financial operations, that channel is an email directed to the payroll manager or department controller, with a subject line that identifies the issue type and the affected run. Some organizations use internal ticketing systems. The critical requirement is that the notification creates a timestamped, searchable record that shows the worker raised the concern before execution. A verbal notification to a supervisor in a hallway does not create that record. If the supervisor later claims the clerk initiated the payroll run without flagging anything, a hallway conversation produces no counter-evidence. A timestamped email to the official reporting address produces a complete one.

The third step is withholding the action pending review. This is the step that most directly exposes the worker to speed pressure, because payroll runs operate on hard deadlines. Direct deposit processing windows close at specific times. Late payroll triggers employee relations consequences and, in states with strict wage payment timing laws, regulatory penalties against the employer. That deadline pressure is real, and supervisors feel it acutely. A clerk who withholds a corrupted payroll run will encounter that pressure directly. Where the clerk's role and company policy support it, the withhold remains the correct action. A payroll run executed with a known, documented error shifts responsibility for the downstream consequences: from the system, onto the person who ran it anyway, despite knowing about the error.

The financial clerk's scenario makes all three steps concrete. A clerk managing weekly payroll for a mid-size operation works from a batch file that an automated payroll platform generates by pulling data from a time-tracking system and an HR database. That pull happens through a system integration, one that requires both platforms to communicate correctly at the moment they sync. When the integration drops a data sync, applies a prior period's overtime rules to the current period, or carries a terminated employee's record forward without deactivation, the resulting batch file looks structurally complete. The format passes every automated validation check the payroll platform runs. The file moves to the clerk's queue for final review and execution.

The clerk's review catches a discrepancy. Three employee records show overtime hours that the time-tracking system's own period summary does not support. Two records show deductions applied at rates that do not match the current benefits schedule on file. The clerk does not execute the run. The first step produces a written record in the department's formal error log: each affected employee ID, the field showing the incorrect value, the value the batch file shows, and the source document that establishes the correct value. The second step produces a timestamped email to the payroll manager and the controller, identifying the affected records, describing the specific discrepancies, and stating that the clerk is withholding the run pending review and correction. The third step holds. The run does not execute until the payroll manager reviews the documentation, confirms the errors, and returns a corrected batch file with written authorization to proceed.

The financial and regulatory exposure that the withhold prevents is quantifiable. The U.S. Department of Labor's Wage and Hour Division enforces the Fair Labor Standards Act. Employers who underpay required overtime can owe the unpaid wages plus an equal amount in liquidated damages. A corrupted payroll run executed under deadline pressure can therefore create direct financial exposure for the employer. The clerk who withholds that run helps reduce the organization's exposure before those liabilities can materialize.

Mandating the Human Anchor

The Human Anchor standard names a specific point that organizations can adopt in any high-risk workflow, where a qualified human being must provide direct, documented authorization before the process moves forward. Once adopted by an organization, the standard treats that point as a binding operational rule, and it applies regardless of what automated approval systems have already processed or confirmed upstream. Binding means the standard holds up no matter the pace pressure. A suggested practice lapses under deadline pressure. A binding standard holds because deviation from it creates a documented procedural failure.

Establishing a binding Human Anchor standard requires four operational decisions. First, the team or department must identify which decision categories carry consequences severe enough to require direct human authorization. Second, it must designate the authorization role by job title so that the standard survives personnel changes. Third, it must define what form the authorization must take: a physical signature, a logged system entry, or a documented verbal confirmation recorded in writing within a fixed time window. Fourth, it must log every authorization event with a timestamp and the authorizing party's name so that the standard remains auditable. Those four decisions must exist in writing for the standard to function as enforceable policy.

Automated systems create the precise conditions that make the Human Anchor standard necessary. Modern workflow platforms route documents, generate approval flags, and advance tasks through processing queues without requiring a human being to open and read the underlying document. An automated approval flag on a purchase order tells a downstream processor that the order cleared a threshold check. It does not tell the processor whether a qualified human reviewed the line items against a verified vendor contract. A digital sign-off on a clinical order confirms that the order moved through the electronic health record system's routing logic. It does not confirm that a licensed clinician read the order, recognized the patient's current condition, and made a judgment call based on current clinical data. Automated routing produces authorization-shaped outputs. The Human Anchor standard requires actual human review to stand behind those outputs before high-consequence action proceeds.

Physical operations illustrate how the standard works on the ground. A crane lift on a construction site requires a lift plan specifying load weight, boom angle, radius, and rigging configuration. A lift plan may arrive as a printed or digital document and may appear complete before anyone at the site has compared it with current physical conditions. Before the lift proceeds, the designated qualified reviewer examines the plan against the actual equipment, load, and site conditions and provides the required authorization. That sign-off is the Human Anchor point. The lift director's review brings in real-time physical judgment the platform simply can't access, and no automated approval flag can substitute for it.

Exercises:

Exercise 1: Draft a Human Anchor Rule

Identify one high-risk decision category in your current role. Choose a decision where an error would result in physical injury, financial loss, or a regulatory violation significant enough to trigger an official investigation. Write the decision category at the top of a clean page in your bound notebook.

Below the category, write the title of the role responsible for authorizing the action. Use the job title, not a person's name. If your organization has multiple qualifying roles for this category, list all of them.

Below the role, write the required authorization form. Specify whether the authorization must take the form of a handwritten signature on the work document, a dated entry in the department log, or a written confirmation from the authorizing role received within a defined time window before execution begins. Choose one form and write it in a single, unambiguous sentence.

Below the authorization form, write the time constraint. Specify how close to execution the authorization must occur to remain valid. A lift plan signed 48 hours before a lift begins on a site where conditions change overnight does not provide the same protection as a sign-off given within two hours of execution. Apply the same logic to your own decision category and write the appropriate time window.

Combine the four elements into a single written draft rule statement, to propose to your supervisor, using this format: "No [action] shall proceed until [role] reviews [specific documentation] and provides [authorization form] within [time window] of execution." Substitute your own workplace terms for the bracketed items. Write the completed rule in plain language your entire team can read and apply without interpretation.

Date the rule statement, sign it, and submit it to your direct supervisor through your organization's official documentation channel. Request written acknowledgment. File the supervisor's acknowledgment with your original draft in your bound notebook.

Exercise 2: File a Formal, Evidence-Backed Informed Refusal Notice

Identify an incoming work document in your current queue that contains a specific, verifiable error. If your queue is clean, retrieve a document from a prior period that contained a discrepancy you caught and corrected. The document must carry a high-risk claim: a quantity, a dosage figure, a load specification, a payroll figure, or a regulatory code that carries physical, financial, or compliance consequences if wrong.

Open your bound error log and record the following in separate columns: the document name or identifier, the date and time of discovery, the specific field or data point carrying the error, the value the document shows, the value your independent source confirms it should show, and the name or identifier of that source.

Draft a formal Informed Refusal Notice on a separate page. The notice must contain five items: your full name and job title, the date and time, the document identifier, a single-sentence statement describing the specific error with the incorrect value and the correct value placed side by side, and a clear statement that you are withholding the associated action pending correction and formal authorization from the designated role, consistent with your job's actual authority, licensing duties, or company procedure.

Address the notice to your direct supervisor and the next level of authority above them. Deliver it through your organization's official communication channel. Do not use informal messaging. Attach a copy of the relevant error log entry as supporting evidence.

Record the time of delivery and the recipient's name in your bound notebook. If your supervisor acknowledges the notice in writing, file that acknowledgment alongside the original notice and the log entry. If your organization has a defined response window and your supervisor does not acknowledge within it, escalate the notice to the next official channel in your organization's reporting structure and log that escalation with a timestamp.

Conclusion

The fear of replacement sits at the center of this book's argument. Many workers who have picked up a work order, a clinical chart, a payroll batch, a route sheet, or a project spec, and wondered whether a human being or an automated system made it, already sense the reality behind that fear. The fear is not irrational. Automated tools really do eliminate roles. Executive mandates can push unverified systems onto frontline workers faster than consent or preparation can keep up. Automated generation tools really do produce high-confidence errors that look just like accurate, verified work. These are facts, not anxiety, and dismissing them as irrational misses the real structural problem entirely.

The structural problem is specific. Organizations that rush automated tools into daily operations, without building in verification requirements, trade a short-term labor cost cut for a long-term operational liability. Institutional memory erodes when automated systems overwrite the built-up, tacit knowledge that experienced workers carry, about site conditions, vendor behavior, equipment quirks, and customer needs. Speed mandates then wear down worker craft, by cutting out the time and space skilled judgment needs. And customer trust breaks down when automated outputs reach the point of service carrying errors that no human ever reviewed before they went out. Those losses can pile up off whatever dashboard first justified the technology mandate: in the operational record frontline workers keep, and in the costs legal and compliance teams spend months untangling.

Together, the nine skills in this system add up to one complete operational stance, built for an era where a document's origin often can't be confirmed at a glance. Workers who build these skills into their daily routines don't need to know whether an incoming document came from a human colleague or an automated platform. They apply the same verification discipline to both, treating every high-consequence claim as unverified by default and most workflow transitions as an opportunity to catch errors before they execute. That posture is the posture of a skilled operator who understands the tools in use, governs them through structured verification, and maintains the human judgment layer that automated systems cannot replicate.

The displacement trap works two ways. It convinces workers that speed compliance equals competence. And it can lead organizations to assume that algorithmic output volume equals institutional intelligence. Neither claim holds under operational scrutiny. Speed without verification produces confident errors. Volume without judgment produces synthetic garbage at scale. Workers who understand this distinction carry an operational advantage that automated tools cannot generate for themselves.

The nine skills exist precisely at that advantage point. They require human attention, pattern recognition, domain knowledge, and accountability. Automated systems, built for smooth text and generation speed, don't replicate those capacities. They simulate the surface of expert output without the structural integrity that expert judgment provides. Workers who maintain the verification layer, build the documentation record, run the refusal sequence, and hold the Human Anchor in place govern the tools from a position only a human being can occupy: direct accountability for what happens when the work actually executes.

Demonstrated competence is the practical resolution to the fear of replacement. Workers who apply these nine skills daily build a verifiable record of operational judgment. That record sets them apart, both from the unverified outputs they govern and from workers who execute those outputs without a second look. That record becomes real professional standing, in any regulatory dispute or performance review where the organization treats automated output volume as a substitute for human expertise. An organization that loses that standing loses the institutional knowledge and error-catching capacity its automated tools depend on, just to operate without causing harm. Human mastery over verification, documentation, and the formal refusal of dangerous instructions remains one essential foundation for deploying automated tools safely.

Also By Renae C. Linde

Think Sharp

Think clearly when systems, pressure, and noise compete.

Ethical Hustle

Build success without surrendering judgment, values, or boundaries.

When We Let Go

What happens when responsibility quietly gives way to retreat.

References

American Nurses Association. (2025). Code of ethics for nurses. (https://codeofethics.ana.org/)

Autio, C., Schwartz, R., Dunietz, J., Jain, S., Stanley, M., Tabassi, E., Hall, P., & Roberts, K. (2024). Artificial intelligence risk management framework: Generative artificial intelligence profile (NIST AI 600-1). National Institute of Standards and Technology. (https://doi.org/10.6028/NIST.AI.600-1)

Bai, X., Nunez, M., & Kalagnanam, J. R. (2012). Managing data quality risk in accounting information systems. Information Systems Research, 23(2), 453–473. (https://doi.org/10.1287/isre.1110.0371)

Components of a security plan, 49 C.F.R. § 172.802.

Exercise of any right afforded by the Act, 29 C.F.R. § 1977.12.

Fair Labor Standards Act of 1938, 29 U.S.C. §§ 201–219.

Hazardous Materials: Security Requirements for Offerors and Transporters of Hazardous Materials, 68 Fed. Reg. 14,510 (Mar. 25, 2003).

Justesen, L., & Plesner, U. (2024). Invisible digi-work: Compensating, connecting, and cleaning in digitalized organizations. Organization Theory, 5. (https://doi.org/10.1177/26317877241235938)

Kalischko, T., & Riedl, R. (2021). Electronic performance monitoring in the digital workplace: Conceptualization, review of effects and moderators, and future research opportunities. Frontiers in Psychology, 12, Article 633031. (https://doi.org/10.3389/fpsyg.2021.633031)

Kleinberg, J., Ludwig, J., Mullainathan, S., & Rambachan, A. (2018). Algorithmic fairness. AEA Papers and Proceedings, 108, 22–27. (https://doi.org/10.1257/pandp.20181018)

Occupational Safety and Health Act of 1970, 29 U.S.C. §§ 651–678.

Pipeline and Hazardous Materials Safety Administration. (2024). Revisions to civil penalty amounts, 2025. Federal Register, 89, 107144.

Requirements for protective systems, 29 C.F.R. § 1926.652.

Shumailov, I., Shumaylov, Z., Zhao, Y., Papernot, N., Anderson, R., & Gal, Y. (2024). AI models collapse when trained on recursively generated data. Nature, 631(8022), 755–759. (https://doi.org/10.1038/s41586-024-07566-y)

Tabassi, E. (2023). Artificial intelligence risk management framework (AI RMF 1.0) (NIST AI 100-1). National Institute of Standards and Technology. (https://doi.org/10.6028/NIST.AI.100-1)

Transportation of hazardous materials; driving and parking rules, 49 C.F.R. pt. 397.

Underpayment, 26 C.F.R. § 1.6664-2.

About the Author

Renae C. Linde writes analytical nonfiction grounded in psychology, culture, and lived experience. After more than two decades in information technology, business analysis, and labor advocacy, she brings a systems-level perspective to questions of identity, power, and social continuity.

Her work is direct, research-informed, and unsentimental. She examines how individuals and communities adapt under pressure, tracing patterns across time rather than relying on nostalgia or abstraction. Across subjects that range from political division to cultural inheritance, she favors structure over slogans and consequence over comfort.

Linde lives and writes in the American Midwest.

www.ingramcontent.com/pod-product-compliance
Lightning Source LLC
LaVergne TN
LVHW010728061126
851857LV00014B/1390
* 9 7 9 8 9 9 3 0 8 3 0 4 9 *